Prelude to the Space Age

The Rocket Societies: 1924-1940

Frank H. Winter

Published for the
National Air and Space Museum
Smithsonian Institution
by the
Smithsonian Institution Press
City of Washington, 1983

Second printing, 1984

Library of Congress Cataloging in Publication Data
Winter, Frank H.
Prelude to the space age.

Bibliography: p.
Includes index.
1. Rocket research—Societies, etc. I. Title.
TL780.W56   621.43'56   81-607883
ISBN 0-87474-963-8   AACR2

Contents

Acknowledgments

Without the unstinting support of Walter J. Boyne, Director, National Air and Space Museum, the realization of this work would not have been possible. Tom Crouch of NASM's staff continually provided encouragement and invaluable advice. Special thanks are also accorded to my special manuscript reviewer, my wife Fe Dulce R. Winter; to Sheila Ellis, Diane Pearson, Debbie Hatfield, and Barbara Pawlowski, who very patiently typed the manuscript; to Kitty Scott, NASM Librarian; to Richard F. Hirsh, Research Fellow, NASM, for his editorial assistance; and especially to Jay Spenser of NASM's Aeronautics Department, for graciously providing his linguistic talents. Invaluable acknowledgements are also due to Lee Saegesser, Archivist, National Aeronautics and Space Administration, for his encouragement; Dr. Eugene M. Emme, former Historian, NASA; to George S. James, Program Coordinator, National Science Foundation, a loyal friend and fellow rocketry historian who continued to give me fresh insights and always the strongest support; Mitchell R. Sharpe, Curator, Alabama Space and Rocket Center; Frederick I. Ordway, III, Energy Research and Development Agency; Herbert Schaefer, Upper Stages Office, NASA Headquarters, whose intimate knowledge and rich collection of the German Rocket Society has been of enormous help and who has likewise brought to light the early Argentinian astronautical group; G. Edward Pendray, without whose gracious help and patience the American Rocket Society story would never have been told; David Lasser, Sam Moskowitz, and Mort Weisinger who likewise filled up significant gaps in the ARS story. Special thanks are also accorded to founder members of the Society: Dr. William Lemkin with whom I had a most enjoyable afternoon chat; Dr. Samuel Lichtenstein, Charles Van Devander, Adolph Fierst; and later members Alfred Africano, Bernard Smith, and Robert C. Truax. Thanks are also given to Mrs. Laurence Manning and Mrs. Harry Bull, widows of the two pioneers. Thanks also to Ernst Loebell, founder of the Cleveland Rocket Society who provided a wealth of data on his group.

Meritt Williamson and Franklin Gates provided first hand memories and documents of the Yale Rocket Society. Some Soviet material was also very graciously provided to me by Dr. Viktor N. Sokolsky of the USSR Academy of Sciences to whom I am most indebted. On the English side an immeasurable debt of gratitude is due to the founder of the British Interplanetary Society, P. E. Cleator, who very carefully read and pre-proofed the BIS Chapter and who graciously answered all of my many questions; and to Eric Burgess and Arthur C. Clarke, who despite incredibly busy schedules afforded me generous portions of their time. For additional Austrian and German developments credit must be extended to Irene Sänger-Bredt, Krafft Ehricke, Karl Pogensee, Rolf Engel, the late Wernher von Braun, Helmut Zoike, and Ernst Khuhon. Edward Peck also deserves special mention for his dual role as a reviewer—whose corrections and suggestions were adopted almost in toto—and from whose own considerable early astronautical collection came new finds. Thanks also to the librarians and staffs of the Library of Congress; the New York Public Library; the Seeley Mudd Library of Princeton University; the Peoria Public Library, Peoria, Illinois; the Cleveland Public Library; numerous other libraries around the country; and the Lenin Library in Moscow. Finally, very special thanks are due to Mrs. Robert H. Goddard who got me started.

Frank H. Winter

Foreword

"Imagine, grown people really thinking that man could fly to the Moon!" Such was an almost universal public reaction to the small groups of enthusiasts in a half-dozen countries who, as early as the 1920s and 30s, believed in the possibility of interplanetary flight.

The dream of space flight was, of course, many centuries old. The fixed points of twinkling stellar lights in the night skies, the Moon, and the wandering lights later revealed as the planets, had long fascinated humanity. The possibility of visiting such unknown places was a natural extrapolation of mankind's basic urge to explore.

Eventually, the work of such brilliant men as Newton, Galileo, and Copernicus, enabled future researchers to draw proper maps of the cosmos. But the unimaginable distances! The total vacuum of space! The unknown hazards of meteorites, the problems of navigation, and the velocity required to travel in a reasonable period of time. Such considerations both staggered and challenged the imagination.

Theorists rejected flights by balloon or projectile on the basis of their impracticality. By 1900 recognition of the principle of reaction (or rocket) propulsion began to be appreciated as the sole means of attaining practical travel through space. British rocketeer A. V. Cleaver once commented that by 1900 some 90% of the *science* required to go to the Moon was known but the engineering and technology base required had not yet been acquired.

By the 1920s, technology was advancing rapidly. Great strides were made in the burgeoning aircraft industry in alloy metals, aerodynamic principles, and light-weight structural design. It was in this conducive climate that the work of the rocket pioneers became known. Their writings had influenced greatly the formation of the interplanetary societies discussed in this work. Their members were surely optimists and romantics, including many science fiction devotees. In fact, the three recognized pioneers of space flight, Tsiolkovsky, Goddard, and Oberth all acknowledged the influence of Jules Verne during their formative years.

The value of collective discussion in these societies promoted self-education and consideration of factors recognized as necessary for space flight. Society members studied such identifiable problems as rocket propulsion, navigation, and life-support systems. Public forums attracted new members—as well as public ridicule.

And so, despite a depressing lack of funds, rocket motor design and tests commenced in the USSR, Austria, Germany, and the United States. The Great Depression heightened their problems, and yet strengthened their zeal, for escape from the day-to-day problems of living was easier when one conceptualized fantastic journeys to the planets and beyond. Slowly progress was achieved, and publications and correspondence crossed national borders until the horror of war intervened, a war that witnessed the emergence of the first modern liquid-fuel rocket.

Frank Winter has documented the formation and development of these rocket societies, and reveals in print many details hitherto unknown. An indefatigable researcher, Mr. Winter has displayed great persistence in seeking out and interviewing surviving members of these societies. He is to be congratulated for the thoroughness and depth of his research and documentation.

His book is significant for two reasons: it provides, for the first time, a prime reference document embracing the activities of all known rocket societies in the period 1924—1940; secondly, it is a warm tribute to those individuals who dedicated themselves with unselfish enthusiasm to the dream of space flight. Their personal sacrifices, frustrations, and disappointments were common and great. Happily, many of the members of such groups have lived to see their dream accomplished far beyond their wildest expectations, and their private satisfaction is easily understood. Mr. Winter's book gives us yet another opportunity to ponder the full meaning of rocket pioneer Robert Goddard's prophetic words: "The dream of yesterday is the hope of today and the reality of tomorrow."

Frederick C. Durant, III
Former Assistant Director, Astronautics
National Air and Space Museum

Abbreviations

AIS		American Interplanetary Society
ARS		American Rocket Society
BIS		British Interplanetary Society
CGIRD (or TsGIRD)	*Tsentral naya Gruppa po Izucheniyu Reaktivnogo Dvizhenia*	Central Group for the Study of Reaction Motion (Jet Propulsion)
CRS		Cleveland Rocket Society
e.V.	*eingetragener Verein*	Registered Society
EVFV	*E.V. [eingetragener Verein] Fortschrittliche Verkehrstechnik*	Registered Society, Progress in Traffic Technics *(or technology)*
GDL	*Gazodinamicheskaya laboratoriya*	Gas Dynamics Laboratory
GfW	*Gesellschaft für Weltraumforschung*	Society for the Exploration of Space
GIRD	*Gruppa po Izucheniyu Reaktivnogo Dvizhenia*	Group for the Study of the Reaction Motion (Jet Propulsion)
ICSC		International Cosmos Science Club
IRKA	*Internationale Raketenfahrt-Kartei*	International Rocket Travel Index (or International Rocket Travel Information Bureau)
LenGIRD	*Leningradskaya Gruppa po Izucheniyu Reaktivnogo Dvizhenia*	Leningrad Group for the Study of Reaction Motion (Jet Propulsion)
MIRAK	*Minimumrakete*	Minimum Rocket
MIS		Manchester Interplanetary Society
MosGIRD	*Moskva Gruppa po Izucheniyu Reaktivnogo Dvizhenia*	Moscow Group for the Study of Reaction Motion (Jet Propulsion)
NYBICSC		New York Branch of the International Cosmos Science Club

OIMS	*Obschestvo po Izucheniyu Mezhplanetnykh Soobshchenii*	Society for the Study of Inter-planetary Communication
OR	*Opytnaya raketa*	Experimental rocket
ORM	*Optynyy Raketnyy Motor*	Experimental Rocket Motor
OSOAVIAKHIM	*Obshchestvo sodeistviya oborone, aviatsionnomu i Khimicheskomu stroitel'stvu SSSR*	Society for Assisting Defense and Aviation and Chemical Construction in the USSR
PRA		Peoria Rocket Association
REP		Robert Esnault-Pelterie
RNII	*Reaktivni Nauchno Isseldovatel'kii Institut*	Scientific Research Institute of Jet Propulsion
SPDACC	*(variation of OSOAVIAKHIM)*	Society for Assisting Defense and Aviation and Chemical Construction in the USSR
TsAGI	*Tsentral'nyi Aerogidrodina micheskii Institut*	Central Aero-Hydrodynamics Institute
Ufa	*Universum Film, AG (Aktiengesellschaft)*	Universal Film Corporation
VfR	*Verein für Raumschiffahrt*	Society for Space Travel or German Rocket Society
V-2	*Vergetungswaffe Zwei*	(Vengeance Weapon Two), A-4 rocket (Aggregate 4)
WRS		Westchester Rocket Society

I

The Rocket Societies

The date 4 October 1957 is generally regarded as the first day of the Space Age. Then, the Russians launched *Sputnik* 1, the world's first artificial satellite. Yet the *idea* of spaceflight has an ancient history that considerably predates 1957, as the venerable tale of Icarus and other legends attest.

Prior to this century, the astronautical literature was largely fanciful in outlook, consisting of both fiction and speculation. The true birth of the Space Age, however, may be said to have begun when applied science entered the picture in the 1920s and 30s. At that time, the cornerstone theoretical works on the subject were written, the word "astronautics" was invented, the first organizations dedicated to the advancement and accomplishment of spaceflight were founded, the first rudimentary liquid fuel rockets were made and flown, and future spaceship designers received their first training.

This book documents the background, creation, and work of the rocketry and astronautical societies that represented the core of the international astronautical movement of the 1920s and 30s, and it relates how the societies and their members helped lay the groundwork for the true beginnings of what may be called "modern astronautics."

These rocket societies of the 20s and 30s were, simply stated, groups of people who associated for a common purpose—the conquest of space. Exactly how they would go about this nobody knew, though the perfection of the liquid-propellant rocket motor, as theoretically shown by Professor Hermann Oberth in his *Die Rakete zu den Planetenraümen* (The Rocket Into Planetary Space) of 1923, was one answer.

Oberth showed mathematically that far higher energy potential was possible with liquid propellants in contrast to the thousand year old solid, gunpowder type rocket fuels. With a liquid system, Oberth reasoned, it was feasible to make a very large *manned* rocket, capable of reaching escape velocity from the earth and continuing on a sustained flight to the Moon or another planet. He designed such a ship, with all the necessary gear for a manned voyage. Oberth's 1923 study itself was a pivotal factor in opening the international space travel movement of the 1920s and 30s.

Yet the Oberth book and others that followed it presented only theories and broad possibilities. None of the societies had any long-range blueprints for a spacecraft or space voyage, with the exception of the British Interplanetary Society which designed a lunar ship just before World War II but which was to be propelled by multi-cellular solid rockets. Generally, the societies believed the solutions to technical problems would fall into place. Those groups that performed rocket experiments were empiricists. In short, the early society members were the most optimistic yet naive of young men and women. Nevertheless, as we shall see, they made important contributions.

The German Rocket Society, for example, conducted numerous successful—and some not so successful—static tests and flights with liquid oxygen/alcohol rockets. Some Soviet rockets, such as the GIRD-X in 1933, were also fueled with liquid oxygen and alcohol; "lox" and gasoline were normally used. The Germans, and also the American and Russian groups, developed the first static test stands for monitoring rocket engine performance; they developed cooling methods for long duration engines, notably the regenerative method in which the fuel is circulated around the engine prior to entering the combustion chamber; and they gained invaluable experience in handling liquid oxygen which has temperatures close to 200 degrees Centigrade (312° F). They also learned how to utilize superlight, super-strength materials for combustion chambers and other critical rocket parts; various techniques of fuel injection, including in the case of Russian rocket societies, the development of pumps, were worked out; and most importantly, the societies developed the rocket team concept from design to launch.

Among some of the more interesting innovations was the Soviet Gas Dynamics Lab (GDL) produced and tested electric rocket motor of 1929 which, because of its infinitestimally small thrust, could only have had space potential. (Independently, the American, Robert H. Goddard, performed successful ion propulsion experiments in 1916—17 but they were not well known.) The first Soviet flight rocket, GIRD-09 in 1933, was also the world's first hybrid rocket, utilizing both a solid and liquid propellant. The Soviets must also be credited with flying the earliest successful liquid-propellant sounding rockets, such as the *Aviavnito* in 1936, and with developing complex sounding rocket instruments. One was a special miniature still camera which automatically took a series of photos during the ascent. The Russian R-06 sounding rocket in 1939 may have marked the first use of wind tunnels for the design of a scientific rocket.

Directly related to manned spaceflight were the first crude biological experiments conducted by the LenGIRD (Leningrad Group for the Study of Reactive Motion), the American Rocket Society and privately, by Wernher von Braun of the German Rocket Society, with Constantine Generales. These tests consisted of subjecting mice or other animals to centrifuge "rides" to determine the effect of high accelerations upon living organisms. The British too made an early step towards manned spaceflight. Members of the British Interplanetary Society built an optical device for peering out a spaceship while spinning. The "Coelostat," a form of stroboscope, was to be installed in the BIS lunar spaceship of 1939.

There were also individual rocket experimenters in the 1920s and 30s but the greatest strides were made through team efforts, by means of rocket societies or other organizations. The American, Robert H. Goddard, was an exception as he achieved tremendous progress in developing the liquid-propellant rocket while working with only a handful of assistants. Paradoxically, he also shunned the societies yet at the same time was revered and sought by them. In assessing Goddard's role in the history of the societies, we have to consider his impact in two fields: space travel and rocketry. Goddard's 1919 Smithsonian paper *A Method of Reaching Extreme Altitudes* was unquestionably very influential to the space travel movement since he was one of the first to describe scientifically flight to the Moon by rocket. On the other hand, his Moon rocket was propelled by relatively inefficient solid fuel based upon his experiments up to that time with "smokeless powders" (nitrocellulose-nitroglycerine). He began experimenting with liquids from 1920 but throughout his life maintained a rigorous secrecy over all of his work and refused to share his findings. From the rocketry or technical point of view, therefore, Goddard had minimum impact.

It was through the pooled efforts of some of the American Rocket Society members, and not the solitary, though brilliant, work of the loner Goddard, that a valuable inroad was made which contributed towards future American space and rocket technology. Specifically, this was the work of the American Rocket Society's Experimental Committee on their test stand upon which the regeneratively-cooled liquid propellant rocket engine of member James H. Wyld was successfully developed. Wyld, with three other ARS members, subsequently formed a private company which became one of the giants in the new aerospace industry, Reaction Motors, Inc., afterwards a division of Thiokol Chemical Corporation.

In the Soviet Union, the space program has an even stronger connection with the various rocket and space travel societies of the 1930s. Apart from several technological priorities, the Soviets can rightly claim that several of their topflight rocket and spacecraft designers, such as Sergei P. Korolev, received their initial education in rocket techniques directly from their experience as leaders of the earlier groups.

The world's first large-scale liquid propellant rocket, which was the forefather of all modern liquid fuel launch vehicles, the German V-2 rocket of World War II, likewise traced its roots back to the societies. Its civilian manager, and afterwards the most famous rocket scientist and leading spaceflight advocate in the world, was Wernher von Braun. Von Braun had learned his rocket fundamentals as a young member of the German Rocket Society, also called the VfR. Von Braun was the first of several gifted early rocketeers from the VfR and other German space travel societies to join the German Army's rocket development program that culminated in the V-2. Following the war, von Braun, with other VfR veterans, came to America and were in the forefront of the early United States space program.

How many other aerospace contributors were intially inspired by the publicity generated by the early societies and the examples of their experiments, it is impossible to say. For certain, the societies through their relentless, often romantic publicizing of space travel via newspaper and magazine articles, lectures, demonstrations, exhibits, radio talks, and films, influenced millions into accepting the possibility, and even the inevitability, of space travel. It was a revolutionary concept in the 1920s and 30s, though the first scientific theories dated to before the turn of the century. The societies exploited all the media at their disposal in educating the public and scientific community in the ideas of the early theorists and experimenters who led the way—the Russian, Konstantin E. Tsiolkovsky; the American, Robert H. Goddard; the Rumanian, Hermann Oberth; and the Frenchman, Robert Esnault-Pelterie. This was the role of the societies as educators and motivators.

It was a difficult role since skepticism abounded, as with any new and revolutionary concept. Much of the lay public at first believed the archiaic notion that spaceships would not work simply because the rocket needed air "to push against" in order to fly. The societies corrected this inaccuracy and taught the public the fundamental principle of rocket motion expressed by Sir Issac Newton's classic Third Law of Motion: "For every action there is an opposite and equal reaction." The spaceship could work, theoretically.

The scientific community had its own biases. At the time, some of them were valid. They realized the meagerly state-of-the-art of rocketry and the enormous technology and knowledge required for such an undertaking. Some feared the unknown dangers of cosmic rays. Some predicted spaceflight was not possible until another century. Others were less hopeful and saw no future in the liquid or chemical rocket—atomic energy, once it was harnessed, seemed the only answer. Most were ashamed to speculate on the possibilities of space travel as it might sully their "scientific" reputations, particularly among their peers. They did not wish to be linked with crackpot ideas and crackpot organizations.

This was part of Goddard's dilemma. The fire of the space travel dream burned within him since childhood. But shy by nature, and rigorously educated as a Doctor of Physics, he deferred, at first, from having any association with the societies and their publicity. A few, more extroverted and imaginative scientists like Dr. Clyde Fisher of New York's Museum of Natural History, or the British born aerodynamicist, Dr. Alexander Klemin, very early supported the space travel societies and their aims.

Gradually—perhaps because the science fiction writers gave way to the engineers among the memberships of the American, and in some respects, the British Interplanetary Society— space travel became more "respectable" with the consequence that more and more scientists were won over. The ceaseless publicity campaigns, backed up, in the case of Americans, by increasingly sophisticated rocket experiments, were working. (The British were forced to refrain from active experimentation because of an antiquated anti-firework building law that included any private rocketry.) The subtle creation of a sort of prespace age space "lobby," or generation which was prepared to accept space flight, was the greatest contribution of the societies. The war, of course, put an abrupt halt to the movement, but it was rekindled again once peace was declared. The societies, some of them surviving the conflict, continued their work and were joined by innumerable newer space societies. But that is another story.

Paradoxically, in the late 1920s and early 30s, when the movement was first getting under way in the face of the severest scientific and public criticism, the movement found its greatest strength in Germany where scientists have always been regarded with the highest respect. Perhaps this was because the German-speaking Professor Oberth had virtually initiated the modern phase of the movement with the publication of his book, *Die Rakete zu den Planetenraümen* (The Rocket to Planetary Space), in 1923, followed by a greatly expanded version, *Wege zur Raumschiffahrt* (Ways to Spaceflight), of 1929. Other pioneering works were likewise published in Germany, such as one of the earliest studies on space navigation, Walter Hohmann's *Die Erreichbarkeit der Himmelskörper* (The Attainability of Celestial Bodies) of 1925, and Hermann Noodung's *Das Problem der Befahrung des Weltraums* (The Problem of Travel in Space) of 1929. The German Rocket Society was also one of the earliest, and in its lifetime, the most prestigious of all the space travel organizations.

Possibly too, the space travel movement had a special appeal to the Germans as an extension of the romantic *Lebensphilosophie* (Philosophy of Life) which pervaded the Weimar Republic. The *Lebensphilosophie*, at its height, glorified technology in reaction to Germany's defeat in World War I and its subsequent economic woes. The favorite motto of the German Rocket Society, "Long Live the Spaceship!" seems to bear witness to this. *Frau im Mond* (Woman on the Moon), the first science fiction, space-oriented movie which utilized scientific advisors—Hermann Oberth and Willy Ley, both of whom were leading members of the German Rocket Society— seemed to especially epitomize the artistic aspects of *Lebensphilosophie*, complete with search-lights dramatically playing on the giant Moonship awaiting its night launch. It is interesting to note that the movie itself gave tremendous impetus to the space travel movement and played an important role in both the German and American Rocket Societies.

Science fiction was, in the beginning, an inseparable and formidable factor in fomenting ideas about spaceflight and in attracting fellow enthusiasts to start space travel clubs. This in essence, was how the American and British Interplanetary Societies began. Perhaps, considering the Depression, this escapist literature and the grand, but almost unreachable goal of actually achieving spaceflight, was the American and British approach to *Lebensphilosophie*.

At this juncture, we should ask what really motivated the people in the rocket societies to join. This fundamental and essential question has been put to virtually all of the pioneers interviewed for this book. The answers may be summed up by dividing the pioneers into four different categories. The first are the men of sweeping ideas, theorists who established the building blocks of the new science and became the movement's leaders. These men of energy and imagination, but not necessarily organizationally inclined, were figures such as Oberth, Tsander of the USSR, and Max Valier of Austria.

The second type were idealists taken up with the new concepts and sometimes carried away with them. The science fiction writers who made up the first wave of members of the American Interplanetary Society may be said to fit into this group. They saw themselves as messiahs of a new technology and a new age. All were innate publicists, even missionaries, eager to spread the new gospel, and they tended to be more philosophically than technically oriented.

The third group, later constituting the majority of 1930s rocketry pioneers, were engineers or inventive men. Space travel may not have even appealed to them so much as the sheer engineering challenge of overcoming complex technical problems that none, or few, had tackled before: developing and perfecting an efficient liquid rocket engine for intercontinental mail rockets and stratospheric vehicles. Perhaps, in their time, these men were the most realistic of all. This group also tended to think of the rocketry and interplanetary societies as "fun". This is a word often used in their dialogue. Comradeship was also part of the formulae. "Enthusiastic" is another word often employed by these types of pioneers to describe each other.

The fourth type of early space travel/rocketry pioneering-society member, were those who possessed characteristics of all of the above. These are the rarest sort of individuals, like Goddard, though he was fiercely independent and possessive of his work. He later did join the American Interplanetary Society (later the American Rocket Society), but mainly because it served his needs as a "listening post".

In addition, there were other individual and personal motivations which are more difficult to delineate. One type of motivation is recalled by the late Alfred Africano who joined the American Interplanetary Society in 1932, and who was co-winner (with the Society) of the REP-Hirsch Astronautical Award in 1936. He had been trained as a mechanical engineer, but during the early 1930s the only work he could find was as a civil engineer. Then he read an article about the American Interplanetary Society and its attempts to fire a Moon rocket, and he felt that this was an area where he could use his education and training. "I figured that I could help them with my engineering," he said, "so that I would be able to help them and they would help me retain the mechanical engineering training!'

Africano was not alone as a professional engineer who joined the ranks of the active rocketry experimenters of the societies in order to keep up his engineering proficiency. In Germany, Wernher von Braun observed that young unemployed or irregularly employed engineers of the early 30s also took up rocketry work at the German Rocket Society's *Raketenflugplatz* to maintain their skills; they also had considerable time on their hands. Perhaps, too, when the Depression finally lifted, rocket planes would be flying and they would be part of the new industry. In the long run, this did happen in some instances. Africano, for one, was named an assistant to Dr. Clarence N. Hickman in 1942 for official rocket defense work with the National Defense Research Committee (NDRC), supervising firing tests for the famed "Bazooka" anti-tank rocket and related assignments. Following the war, he resumed work with liquid propellant rockets at the Curtiss-Wright Corp. and afterwards with Chrysler Corporation's Missile Division where he contributed to the Redstone project. Later, he directly assisted the Apollo Moon landing program, working from 1962 to 1970 at North American Rockwell's Space Division; Africano even contributed to some preliminary investigations for the Space Shuttle before he retired.

Africano also recalled that G. Edward Pendray, one of the founders of the American Interplanetary Society, once made a study of his fellow members to determine what motivated them to join. Pendray concluded "that the one thing that most of our members have in common, whether they're attorneys or writers, science fiction writers, or engineers, or what not, is that they have imagination. They have a freedom of thinking or imagination that very few people have."[1]

We should also note, however, that there have been many aerospace developments and pioneers who were not directly linked to the societies; though some were inspired by their work. Dr. Frank Malina, for one, who became a leader in the development of the large scale liquid propellant sounding rocket, had closely followed the work of the ARS, but independently of the Society, he helped initiate a professional research team in 1936 at the California Institute of Technology. Elsewhere, the late Val Cleaver, who became one of Britain's leading rocket engineers from the 1950s until his death in 1977, had been a devoted member of the British Interplanetary Society but really began his professional rocketry career in private industry.

The societies did not immediately precipitate "our" Space Age, starting with the first Sputnik and Explorer artificial satellites during 1957—58, but they did pave the way in training some of the men who launched those first satellites and spaceships and conditioned the world into preparing for the Space Age.

The stories of all the societies is presented here in chronological sequence in order of the respective founding dates, up until 1940. With the subsiding of the Great Depression and throughout the war, the space travel movement took a wholly different turn. It was subdued, and in some countries suppressed, because the same rockets that could drive spaceships could also propel offensive weapons. Besides these considerations, many in the young idealist wave that had started the societies had largely dispersed by 1940. In some countries, a few individuals had already been hired by the military for "secret" rocket work. The war itself took its toll on the members and some of the smaller societies. The present study confines itself to the earlier formative years of the societies before 1940. They were years intense with promise and activity—the prelude to the Space Age.

II

The Background

The Moon Stories—Lucian to Jules Verne

If the foundations of the Space Age are placed in the 1920s and 30s, then its "pre-history" may be subdivided into two distinctive earlier periods. The first can be dated before 1877 and the second after 1877. In that year astronomer Giovanni Schiaparelli discovered Martian "canals."

Before 1877 concepts of space travel and life upon other worlds were fantasies or theological and philosophical speculations. Most often they were combinations of all three. The genre of space travel literature was begun in a satirical-philosophical vein. Lucian of Samosota's *Vera Historia* (True History) of ca. 160 A. D. was an entertaining satire in the purest tradition and may be compared with *Gulliver's Travels* and *Baron Munchausen:* tall tales with witty morals and certainly not to be taken as "true history." Lucian, the Greek satirist, was poking fun at the outrageous fictions put forth as the truth by his past and contemporary poets and historians. His vehicle for travelling to the Moon was incidental to the adventures he wove for his travellers: a ship lifted to the Moon by a terrible whirlwind. In Lucian's second space travel story, *Icaro-Menippus,* the trip to the Moon is planned. Menippus, the hero, observes the night sky for a long time and finally sets off from the summit on Mount Olympus with the help of two wings fastened to himself, one of a vulture and the other of an eagle.

Three centuries later, the great Italian poet Lodovico Ariosto also conceived a Moon flight. His epic poem *Orlando Furioso* (The Madness of Roland), written in 1516, tells of the search by Astolpho for the lost mind of Orlando. Astolpho goes as far as the Moon in his quest, and he travels in a chariot drawn by four red horses. In Germany, the astronomer Johannes Kepler's *Somnium* (Dream), posthumously published in 1634, transports the reader to the Moon through a dream. His tale is couched in supernatural terms to avoid religious and political censure for his own concepts of probable conditions on another world. In England, the Bishop of Hereford, Francis Godwin, also cloaked himself against censure for his Moon travel tale, *The Man in the Moon: Or A Discourse of A Voyage Thither.* It was published in 1638 under a pseudonym, Domingo Gonsales, "The Speedy Messenger." Gonsales flies to the Moon by 30 or 40 trained wild swans—ganses—who pull a chair.

In the same year Godwin's book appeared (1638) was published the work of another English bishop, Bishop John Wilkins of Chester's *The Discovery of the Man in the Moon: or, A Discourse Tending to Prove, that 'tis probable there may be another habitable World in that Planet.* Wilkins tells no fantastic stories but presents a straight-forward speculation on the probable character of the Moon and of its conditions for habitability by earthmen and possible lunarians. We are not told of the means of propulsion through and past the "orb of thick vaporous air that encompass the earth." In France, Cyrano de Bergerac's two space novels, *Voyage dans la Lune* (Voyage in the Moon), of 1649, and *Historie des États et Empires du Soleil* (History of the States and Empires of the Sun), of 1652, were written solely for fun. They were romantic tales in which the dashing heroes succeed in lifting themselves in several ways: by vials filled with dew that mystically rise with the rays of the morning sun; by pieces of lodestone continuously thrown up and therefore pulling a car; by greasing oneself with marrow as the Moon is known to "suck up the Marrow of Animals"; and—quite unintentionally—by a more surer way, through tiers of skyrockets mounted on a special box.[2]

And so the literature runs, all the way down to Jules Verne's 1865 and 1870 classics, *De la Terre á la Lune* (From the Earth to the Moon), and its sequel *Autour de la Lune* (Around the Moon). These latter works were true milestones, both in science fiction and astronautics. The stories were written with the utmost care to be scientifically *possible,* or at least *plausible.* Verne consulted more than 500 reference books to get his "facts" straight and he also gained much from numerous discussions with his cousin, Henri Garcet, a mathematics professor of the Lycée Henri IV in Paris. Verne made his share of technical mistakes but his minutely detailed descriptions of take-off, acceleration (unfortunately, from an inaccurately conceived giant cannon shell), weightlessness, the appearance of the Moon, general conditions in outer space, and rockets for softening the Moon landing and as navigational means were vivid enough to have inspired almost all of the leading astronautical pioneers of the next century—including Tsiolkovsky, Esnault-Pelterie, Oberth and others.[3]

The Mars Period—Schiaparelli to Burroughs

The year 1877 was a turning point and the beginning of the second period of the prehistory of the Space Age. It was in that year that the Italian astronomer Giovanni Schiaparelli made his famous sightings, incorrectly as we now know, of "canals" on the planet Mars. Schiaparelli came to believe his own mistranslation of *cannali,* or natural "channels" of water. The translation of that single word from the Italian into the incorrect English rendering of "canals" considerably altered not only Schiaparelli's but man's view of the heavens. Canals were assumed to be the work of intelligent beings. The immediate and startling conclusion to this premise was that Mars was inhabited by intelligent life forms. The Schiaparelli sightings were revolutionary in several respects. They came not from the clergy nor the romanticists but from the world of science, from a respected Italian astronomer utilizing some of the finest instruments available. However distorted Schiaparelli's monumental discovery, a large segment of the scientific community was convinced that extraterrestrial life existed and that man's destiny was to communicate with them if not to explore Mars and other planets.

In France, the greatest proponent of the habitability of Mars was the balloonist-astronomer Camille Flammarion. Flammarion had delved into the possibilities of extraterrestrial life far earlier. In 1862 he produced his first works on the subject, *Les Habitants de l'Autre Monde* (The Inhabitants of Another World), and *La Pluralité des Mondés Habités* (The Plurality of Inhabited Worlds). From 1877, however, his concentration was clearly fixed upon the red planet. That year saw the publication of his *Stella,* a metaphysical-astronomical tale of a dead man and woman reincarnated on Mars; *Cartes de la Lune et de la Planeté Mars* (Maps of the Moon and of the Planet Mars); and *Les Terres du Ciel* (The Worlds of the Sky), an armchair travelogue of the planets, with particular attention to Mars. Flammarion's greatest contributions to the Mars mystique was his two-volume *La Paneté Mars et ses Conditions d'Habitabilité* (The Planet Mars and Its Conditions of Habitatability), 1892 and 1909. In America, Percival Lowell's name was synonymous with Martian "canals." The Lowell Observatory in Arizona was built for the purpose of studying the canals. Lowell also wrote several notable works, including *Mars as the Abode of Life* (1910).

By the dawn of the new century, two camps had been formed, the canalists and the anti-canalists. The astronomical-biological issues are irrelevant here. The crux of the matter is that the controversy had been started and that men of science and all the dreamers of the world looked towards the heavens as never before. Men of the stature of Nikola Tesla and Guglielmo Marconi seriously began to devise means of communication with the Martians. These were either by powerful light beams aimed by reflectors, or by radio waves. Amherst Professor David Todd, in conjunction with the New England Aero Club, had another novel plan. In May 1909, the Club offered Todd their balloon so that he could receive Martian wireless signals with airborne antennae.[4]

The Age of Spaceship was yet to come but was more than amply presaged by the spaceship and trips to and from Mars in fiction. There is a related and useful survey of science fiction stories in George Locke's bibliographic *Voyages in Space* (1975) in which Mars clearly comes out ahead. Locke's survey covers this genre of literature from 1801 to 1914. On his time chart there is a quantum leap of output of "interplanetary fiction" from the 1870s, with a high plateau reached in the mid-1890s. Among the most famous of the Martian stories are Kurd Lasswitz' *Auf Zwei Planeten* (On Two Planets) (1897) and H. G. Wells' *War of the Worlds* (1898).

Within our own century the pace was set by Edgar Rice Burroughs who launched an entire series of novels with Martian settings with a six-part serial in the pulp magazine *All-Story* in 1912. "Under the Moons of Mars," afterwards retitled *A Princess of Mars* when it came out as a book, was followed by such sequels as *The Gods of Mars* (1918), *The Warlord of Mars* (1919), and many others. Although the grand master of science fiction, Jules Verne founded the art, the genre does not appear to have fully blossomed until the post-Schiaparelli period. Science fiction and scientific speculation, more than the rise in technology itself, seem to have been the real impetus towards the growing astronautical movement. By the 1920s it finally culminated in the founding of the astronautical societies.[5]

Gropings in Applied Science

Progress on all technological fronts since the turn of the century was certainly not without its own impact. This was especially true in the world of aeronautics. A search through the aeronautical patent literature shows an occasional rocket or other reaction-propelled machine vying with the ornithopters, muscle-powered, screw-driven, and other flying apparatuses.

By the mid-1890s, Wilhelm Maybach's carburetor and Gottlieb Daimler's internal combustion engine temporarily put the rocket and other reactive means out of competition as a potential flying machine power plant. Nonetheless, following the realization of heavier-than-air manned flight in 1903 by the Wright brothers, flight beyond the atmosphere was taken out of the realm of the impossible by many.

In Russia, the same epochal year the Wrights flew above the dunes of Kitty Hawk, the deaf provincial school-teacher Konstantin E. Tsiolkovsky published his classic article *Isseldovanie mirovykh prostranstv reaktivnymi''* (Exploration of Cosmic Space by Reaction Apparatuses). Earlier, in Worcester, Massachusetts, the schoolboy Robert H. Goddard had dreamed of a trip to Mars after reading Wells' *War of the Worlds*. By the time of the Wrights' flight, young Goddard was determined to make a career of the study of "reactive motion." And in 1912 in France, the famed early *aeronautical pioneer* Robert Esnault-Pelterie presented to the *Société Française de Physique* his first theoretical concepts of interplanetary flight. Thus, before the beginning of World War I, the seeds of the astronautical movement had already been sown.[6]

The 26 April 1913 issue of *The Scientific American Supplement* published a brief article indicative of the trend, entitled: "Travelling Through Interstellar Space—What Type of Motor Would You Employ?" Allusion was made to the foolhardy and pointless manned skyrocket flight of the dare-devil Frederick Rodman Law in March of that year at Jersey City; Law who was a stuntman and not a space travel enthusiast, miracuously survived the explosion of his giant rocket though hardly caused an *éclat* in scientific circles. The *Scientific American* mainly spoke in more sober, theoretical tones of space flight and extracted much of its data from an earlier and lengthier piece appearing in its counterpart French journal, *Cosmos* for 16 January 1913. Both the piece in the *Scientific American* and the French article entitled, *"Un moteur pour aller de la Terre à la Lune,"* dismissed the airplane and proceeded at once to consider Robert Esnault-Pelterie's *"moteur à réaction"*, the rocket. His estimate of 21 million foot-pounds of energy required for his Moon rocket was, in retrospect, an incorrect approach but certainly reasonable in 1913. He had also considered atomic energy in the form of radium. This later suggestion was derived from another source, a remarkable Belgian patent (No. 23,6377) by Doctor André Bing, granted 10 June 1911. Bing, in effect, patented a spaceship, apparently the first such specification anywhere. He conceived a multi-stage rocket utilizing either solids, gases, or liquids and also nuclear fuel. Hence, it was left to the famed French airplane manufacturer Robert Esnault-Pelterie to pick up an obscure Belgian patent, investigate its astounding propositions mathematically and later to publicize them. (REP, as he was known, was not to publish his complete findings until his 1928 *L'Exploration* and his 1930 *L'Astronautique*). Publicity, these early pioneers were to discover, was everything. Whether their methods were technically sound was immaterial. What counted was that by World War I the spirit of inquiry was there, both as to *why* we should explore the planets and to *what* the possible means of propulsion might be.[7]

Work of the Pioneers—Goddard, Oberth, and Tsiolkovsky

Still, not until after World War I did the astronautical movement gather full momentum. By 1930 it was in full flower and ripe for attracting cadres of people of like interests to astronautical societies. The primary catalyst was not necessarily the war itself with its startling display of the potential of the airplane, but the works of the great pioneers.

Goddard stands pre-eminent not only on the American scene but internationally. This does not mean that his work was understood or was fully mature. On the contrary, Goddard's earliest researches were wildly distorted beyond recognition by the popular press and were not even fully comprehended by the rocket community of the late 20s and early 30s. The fault, if indeed it was a fault, for this state of affairs was Goddard's own. His penchant for secrecy set him apart from the mainstream and implanted erroneous ideas into the heads of the scientific and the lay communities alike. The single publication that, paradoxically, established his reputation as both the foremost rocketeer and as an eccentric "Moon professor" was his 1919 Smithsonian paper, *A Method of Reaching Extreme Altitudes*. The story of his hypothetical Moon rocket exploding flash powder on the lunar surface is well known. Goddard dared use this supposition to argue the ultimate potential of the rocket, to show how much propellant was necessary to achieve the task and how much flashpowder payload was required given the present state of technology. Many readers took his example wholly out of context and transformed him overnight into something bordering upon a crackpot. Goddard, characteristically shy by nature, was never quite the same after this blow. He was thus the spiritual leader of the new movement, but remained aloof from the astronautical societies who hungered for his endorsement and he refused all details of his work. As a result, Goddard's monumental advances in liquid-fuel technology were largely unknown until as late as 1936 when his second Smithsonian report, *Liquid Propellant Rocket Development,* appeared.[8]

Of greater direct influence to the astronautical and rocketry movements was Professor Hermann Oberth of Transylvania, Rumania, who later became a German citizen. Oberth's *Die Rakete zu den Planetenaümen* (The Rocket Into Planetary Space) of 1923 was the pivotal work. Like Goddard's monograph, it too gave mathematical form to spaceflight but it went far beyond. Goddard's 1919 paper dwelt upon conventional solid propellants and suggested an unmanned trip to the Moon in such terms as to make it seem entirely practicable in 1919. Oberth's spaceships were liquid-fueled and his technology a promise of what ultimately lay ahead in manned spaceflight. Goddard has been far too cautious. He had barely mentioned liquids in a 1914 patent and in the 1919 paper, and had only privately been thinking about manned spaceflight. Oberth boldly gave the lay, and especially the scientific community, what it really wished to hear—word of man's involvement in space travel. Above all, he made spaceflight an engineering problem to be worked upon. Oberth considered, for example, the hitherto largely unexamined problems of space food, space suits, space walks, probable missions (land surveying, etc.), and all the minutia in both operating an orbital space station and embarking upon long distance interplanetary journeys.[9]

Beyond the milestone set by the publication of Oberth's book were a host of other pioneering efforts, ranging from the popular *Der Vorstoss in den Weltraum, eine technische Möglichkeit* (Advance into Interplanetary Space, a Technical Possibility), by Max Valier in 1924, to the more erudite and scientific *Die Erreichbarkeit der Himmelskörper* (The Attainability of Celestial Bodies), by Walter Hohmann in 1925.[10]

This background of the astronautical societies would be incomplete without relating the earliest Russian developments. The name Konstantin Eduardovich Tsiolkovsky early looms above all the others There is no question that his writings anticipated the development of astronautics by many decades and that he made innumerable and worthy contributions to the early space travel literature. The Russians date his first concepts of interplanetary travel to 1883. He is credited with deducing some of the fundamental laws of reactive motion in outer space and proposing the liquid oxygen/liquid hydrogen rocket in his 1903 article, "The Investigation of Universal Space By Means of Reactive Devices" which appeared in the journal *Nauchnoe Obozrenie.* Part two of this work, appearing in *Vestnik Vozdukhoplavaniya* in 1911 and 1912 (Nos. 19-22 and 2-9, respectively), considerably expanded his theories. Even earlier than Oberth, he considered space suits and space food, or what Tsiolkovsky called "nourishment and respiration." In short, the Russian did not confine his concepts to launch dynamics and orbital mechanics. He was a significant pioneer in what is known today as life-support systems.

By directly coming to grips with the element of man and the sustaining of his environments in outer space, Tsiolkovsky truly deserves the sobriquet of "Father of Cosmonautics." In the final analysis, the objective historian must seek the full measure of his impact. A thorough search through the astronautical literature of the 1920s and 30s, as well as of contemporary private rocketry—astronautical correspondence (such as the Robert H. Goddard, Willy Ley, and G. Edward Pendray papers), reveals the answer. Details of K. E. Tsiolkovsky's works were essentially unknown outside his native country. His name is not prominent at all in these letters and appears not once in the three-volume *Papers of Robert H. Goddard.* Further, Tsiolkovsky is afforded five scant pages in Werner Brügel's *Männer der Rakete* (Men of the Rocket) (1933), a collection of autobiographical articles on the leading rocket pioneers of the day. Nor do bibliographies of the period show any translations of his works. None of his technical articles appear in the open literature of the West, apart from very generalized pieces by his compatriot Alexander B. Schershevsky appearing in the late 20s. Poverty of finances and modesty dictated part of this situation rather than secrecy. M. K. Tikhonravov, the editor of his collected works, says as much when he writes: "We should say a few words concerning the distribution of Tsiolkovsky's works on rockets. His basic works were published in the periodic [Russian] press during 1903—1912 in quantities of approximately 4,000 copies. The Kaluga editions of Tsiolkovsky himself were usually published in lots of 2,000 copies and were distributed entirely by him. If we add to this the Selected Works, we find that there were only 7,500 copies of his writings. For a period of 30 years this is a very insignificant number."[11]

Thus while the Russian astronautical patriarch Tsiolkovsky appears to have had a clear historical priority, by default his works had no real impact in the West during the period 1903—1923. This is not to deny his true greatness, as Goddard should likewise be forever remembered as being far ahead of his time and brooking no peers. Certainly in his own country Tsiolkovsky was, and continues to be, the spiritual head of the astronautical movement. As such he assuredly stimulated the formation of the first Russian groups, though as an impoverished deaf provincial school teacher he was not able to play a leading organizational part.[12]

The Publicists

Publicity in the astronautical movement of the 1920s and 30s was extremely important, as we have stated. In Russia the man who was in the forefront both as a publicist and a highly gifted and original astronautical thinker in his own right was Fridrikh Arturovich Tsander. It was he, not Tsiolkovsky, who was the leading light in the first Soviet astronautical societies. The Latvian-born Tsander was a devoted student of Tsiolkovsky. He had been introduced to the visions of the old pioneer during his last year at Riga High School in 1902 when he read Tsiolkovsky's "The Investigation of Universal Space." The Mars fever had also captivated Tsander. His biographer suggests the planet became an obsession. A lifetime slogan was "Forward to Mars!" Despite difficult times in Russia, both economically and politically, Tsander actively lectured on the cause of interplanetary flight from World War I until the end of his short, brilliant life in 1933 when he died at age 46. Lenin himself was at one of these meetings, the Provincial Conference of Inventors in Moscow in 1920. "After the speech," Tsander recalled: "I was invited to meet Lenin; this made me confused. Lenin was greatly interested in my work and my plans for the future; he spoke with such simplicity and cordiality that I am afraid I took advantage of his time by relating to him in great detail my work and my determination to build a rocket spaceship . . . At the end of our conversation, Lenin shook my hand strongly, wished me success in my work, and promised support."[13]

We will never know exactly what transpired behind the political curtains of the Soviet Union in regard to any promised support by the Government. Initially, it seems to have been a sincere but unfulfilled pledge. Was Lenin really interested in space travel? This is another supremely enticing question that might never be adequately answered. The idea of space flight itself, in the early 1920s, was revolutionary. It called for far-sighted makers of new worlds, and Stalin's elevation of Tsiolkovsky in later years as a member of the Party and Tsiolkovsky's recognition by the Soviet public as a "Socialist pioneer" bears witness to this. Coupled with political ideology also was an underlying nationalism. Tsander himself, after a speech made in the Great Physics Auditorium at the Institute of Moscow on 4 October 1924, responded to the question "Why do you want to go to Mars?": "Because it has an atmosphere and ability to support life. Mars is also considered a red star and this is the emblem of our great Soviet Army." By the 1930s the Soviet propensity for establishing "firsts" was also at stake. In this, it was a justified case of Konstantin Tsiolkovsky, the Soviet "Father of Cosmonautics" vs. the priority claims of "the German" Hermann Oberth and "the American" Robert Goddard.[14]

The reputation of Tsander also attracted Lenin. Tsiolkovsky was the prolific writer, Tsander the chief spokesman. With the drive and conviction of a crusader, which indeed he was, Tsander lectured wherever he could and in the end probably burned himself out. At the same time, Tsander was performing another service to the growing astronautical movement. He was spreading the names and accomplishments of Tsiolkovsky, Goddard, Esnault-Pelterie, and later Oberth, before the Russian public.[15]

Within Russia there had likewise been an explosion of "scientific fiction" literature from the latter half of the 19th century as there had been in the West. Alexei Tolstoy wrote a Mars story, *Aelita,* in 1923, later made into one of the first Soviet science fiction silent films (1924). Nor were the tales of Jules Verne, Edgar Rice Burroughs, and particularly H. G. Wells, without their vast followings in the Soviet Union. All of these works were quickly translated. After the Russian Revolution, the utopian theme was especially exploited. A Bogdanov's political-interplanetary novel *Krasnoya zvezda* (The Red Star) of 1908 appeared in at least six more editions after 1923. Also of immeasurable importance in promoting the dream of space flight was the first great Soviet "popularizer" of the subject, the physics teacher Yakov Isidorovich Perelman. He was active on three fronts: educating the public, especially the youth, on the scientific possibilities of interplanetary flight, notably through his *Mezhplanetnye Puteshtviya* (Interplanetary Travel) which went through ten printings from 1915 to 1935; writing his own widely-read science fiction; and actively taking a part in the formation of one of the first Russian astronautical societies (LenGIRD). Nikolai Alexyevich Rynin, a distinguished aeronautical professor, former balloonist, and writer, became the leading second-generation astronautical publicist in the USSR. He too was also intimately involved in the creation of LenGIRD. Rynin's writings were extensive, notably his nine-volume encyclopedic masterpiece *Mezhplanetnyie soobshcheniya* (Interplanetary Flight), published from 1928—1932 when the movement was already in full flower.[16]

In the West, the foremost first-generation publicist who was similarly a moving force in the organizations, in this case the Austrian groups and the German Rocket Society, was Max Valier. Valier's career was as intense as that of the Russian Tsander. He ended as one of rocketdom's first fatalities when he was killed in 1930 while experimenting with a liquid-propellant engine for a rocket car. His *Der Vorstoss in den Weltraum* (1924) has already been cited. Long before this, like Tsander, he had been on the lecture circuit and had already produced an impressive list of credits. Valier's biographer traces his first thoughts of rockets and space travel back to his school days at Innsbruck University in 1913— 1914. In 1914 he also produced his first literary work, a fanciful operetta, *"Die Mondfee"* ("The Moon Fairy"). The majority of his written works thereafter were mainly astronomical and metaphysical in nature. Still, his reputation as an authority on space travel had been clearly established by the time of his first major astronautical work in 1924. Valier's role as chief rocket publicist in western Europe of the early to late 20s was subsequently assumed by Willy Ley who, from the appearance of his first book, *Die Fahrt ins Weltall* (Travel in Space) in 1926, was the preeminent astronautical author-lecturer for 40 years. Ley will likewise be encountered in the organizational story of the German Rocket Society.[17]

One more name on the German scene should not be overlooked: Herman Ganswindt. His story is an anomaly. As early as 1891 the eccentric and inventive Ganswindt proposed a spaceship propelled by dynamite cartridges exploding inside a chamber against its walls, thereby, as he thought, driving the ship to other planets by kinetic energy! Distorted as his physics were, he was thinking of a kind of reaction propulsion adapted for space flight. Moreover, his presentation of the idea was made public in the form of a lecture on 27 May 1891 at Berlin's Philharmonie hall. Newspapers of that time reviewed the talk and his "plans" in general as did such periodicals as *Die Zeit* of Vienna, for 28 July 1900. The plans were never fulfilled and the concept largely forgotten until the formation of the German Rocket Society more than three decades later. It is difficult to assess Ganswindt's overall impact upon the astronautical movement, both of the 1890s and during its heyday in the late 1920s and early 30s. In retrospect, if the movement had missed him entirely, it would have been short one of its more colorful individualists. His story is related in more detail below.[18]

Finally, we return to America. The publicists there were relative late-comers. True enough, an enormous amount of interest had been generated by Percival Lowell and his fellow canalists, and by the work of Goddard. But there were really no counterparts to Valier, Ley, Perelman, and Rynin in the Western Hemisphere. Not until 1931, a year after the formation of the American Interplanetary Society, did the first American astronautical book appear. David Lasser, founder of the Society, wrote it. *The Conquest of Space,* however, was solely for popular consumption and paled by comparison to the achievements of the Europeans. Lasser stayed with the Society briefly before embarking upon a totally new career in trade unionism, but he did produce a number of popular articles. G. Edward Pendray then became the leading light, both in Society affairs and as the most widely known and read American authority on rocketry and space flight other than Goddard.

Even before this second generation of publicists, there resided in New York Hugo Gernsback, a native of Luxemburg. Gernsback was much more than the founder of the first science fiction magazine and the world's first radio magazine. He was also a highly effective promotor of the space travel idea. This was accomplished through occasional interplanetary stories in his publications *Modern Electrics,* and *Science and Invention.* These journals culminated in the specialized "scientific fiction" magazine, *Amazing Stories,* which first appeared in April 1926. *Amazing* and later Gernsback magazines habitually ran space travel stories. The gaudy covers painted by Frank R. Paul, the Austrian-born artist, played their own subliminal magic in impressing upon a generation or two the lure as well as the unspeakable dangers of outer space. Yet, Gernsback's contribution to the movement hardly ended there. He himself, and his *Science Wonder Stories* staff of 1930 were directly responsible for launching the American Interplanetary

Society. More than any other nation, America traces its astronautical roots to a science fiction fatherhood.[19]

The word "astronautics" itself was created by a science-fiction author. The term was invented by the Belgian writer J. J. Rosny, *ainé* (the elder), a pseudonym for Joseph-Henri-Honoré Boex. On 26 December 1927, Rosny, with friends Robert Esnault-Pelterie, André Louis-Hirsch, and others, gathered in the home of Hirsch's mother at 47 Avenue d'Iena in Paris to form a committee with the *Société Astronomique de France* to promote space flight. At the end of dinner, they adopted a plan for what became known as the REP-Hirsch Prize. A word was needed to describe the subject of the prize. Rosny thought of *astronautics,* an almost literary invention, meaning "navigating the stars." Esnault-Pelterie inaugurated the word thereafter in his space travel talks and in print. It became the title of his most famous work, *L'Astronautique* (1930). Thus, by 1930, "astronautics" was legitimized. It was both a new entry in the languages of the world and a body of literature on the ultimate realization of the still youthful science of space flight.[20]

III

The Forerunners

First American and Russian Groups

The same two nations that fulfilled the dream of spaceflight, Russia and America, were also the earliest to form organizations dedicated to this goal. However, the first efforts of Russia and the United States were hardly in the nature of a clash of national prestiges. They were conducted under far more innocent circumstances and on a private basis. They were also obscure and unsuccessful, but they were also harbingers of far greater and more successful efforts to come.

The first American rocketry organization is shrouded in obscurity. It was apparently a one-man affair known as the Rocket Society of the American Academy of Sciences and was established in Savannah, Georgia. The founder and first president was a man with the decidedly Slavic name of Dr. Matho Mietk-Liuba. The date was 1918, but little else is known. Dr. Liuba was interested in rocket research since 1915, though in what capacity is not said. He later appears to have left Georgia for New York City. Nothing is heard of him in the early journals of the American Interplanetary Society (later the American Rocket Society), which functioned in the New York City area from 1930. Nontheless, by 1937 Dr. Liuba was still active and his "Society" was apparently a "going affair," as it then reportedly merged into the American Academy of Sciences.[21]

The first Russian group was more substantial in terms of membership and program but only lasted a year. In effect, there were two groups, one emerging from the other. The former group grew out of a lecture delivered by Fridrikh Tsander on 20 January 1924, before the Technical Section of the Moscow Society of Amateur Astronomers. After summarizing his article "Flights to Other Planets", which appeared in the magazine *Samolet,* and in which he presented estimates of energy and velocities required for trips to Mars and Venus, Tsander also proposed the organizing within the USSR, of a "Society for the Study of Interplanetary Travel." The motion was upheld. In the middle of April the group was formally constituted as part of the Military Science Division of the N.E. Zhukovsky Air Force Academy in Moscow and was known variously as the Interplanetary Communications Section or the Jet (i.e., "Reaction Engine") Section. The Secretary of the new group, M. G. Leiteisen, at once tried to enlist the support of Tsiolkovsky. A letter was sent to Kaluga bestowing praises of the highest esteem upon the old pioneer and inviting him to participate in the group's program. Tsiolkovsky replied on 29 April: "Dear Comrades, I am happy to open a section of the Interplanetary Society concerning trips and lectures (by me). But I can promise very little. If I were younger and healthier, I would better able to fulfill your wishes . . ." With or without Konstantin Tsiolkovsky's active participation, the organization would proceed towards its momentous goals. Its objectives were:

1. To bring together all persons in the Soviet Union working on the problem.
2. To obtain as soon as possible full information on the progress made in the West.
3. To disseminate and publish correct information about the current position of interplanetary flight.
4. To engage in independent research and to study in particular the military application of rockets.

This program was refined somewhat and demilitarized when the section was reorganized in May into the successor group known as the Society for the Study of Interplanetary Communication, or *"Obschestvo po Izutcheniyu Mezhplanetnykh Soobschenii"* (OIMS). Its aims were basically the same.[22]

News of "breakthroughs" on questions of interplanetary flight and of the formation of the section later appeared in the December 1924 issue of *Tekhnika i Zhizna'* (Technology and Life), a Soviet magazine roughly akin to Hugo Gernsback's *Science and Invention.* "For its members," the article says, "the section has been provided with a number of lectures—including those by Professor [Vladimir P.] Vetchinkin and Engineer Tsander; a competition was announced for designing a small rocket to travel 100 versts [68 miles or 110 kms]; a club was formed for a more comprehensive theoretical study of the question; a laboratory was organized, a book-stall was opened for satisfying the extensive demand for literature, a movie group was set up which is now developing scenarios, and so forth. The section is participating actively in the organization of the Society for Interplanetary Travel. The first step taken by the Society was to arrange for a public lecture by M. I. Lapirov-Skobolo at the Polytechnic Museum in Moscow. The tremendous success attained with the lecture provides evidence as to how high the interest is in the question of space travel. Plans call for publishing on 1 July the first issue of the journal *Raketa,* an organ of the Society. The Society is temporarily quartered at the observatory of Tryndin (Moscow, B.

Lubyanka, No. 13)." *Raketa* was never published. Rudolf Smits' *Half A Century of Soviet Serials 1917—1968,* reveals only that the earliest Russian magazine bearing that name was *Raketnia Tekhnika,* a Russian translation of the *Journal of the American Rocket Society* from 1961.[23]

The original OIMS group comprised about 25 people. Following its reorganization, and particularly after the Lapirov-Skobolo speech "Interplanetary Travel," membership swelled to almost 150. A more definitive break-down of the membership of OIMS as recounted by its Chairman, Kramarov, in an interview with the writer Evgeny Riabchikov, was as follows: 53 students, 43 workers and white-collar employees, 14 "science technicians," 6 journalists, and 5 scientists and inventors, for a total of 104 men and 17 women. OIMS Chairman Grigory Moiseyevich Kramarov was a 37-year old writer and member of the Communist Party since 1907. Moris Gavrilovich Leiteisen was Secretary. Other key officers were Valentin Ivanovich Chernov, V. P. Kapersky, M. A. Rezunov, M. G. Sererennikov, and of course, F. A. Tsander.[24]

Eager to see as early a start as possible on research related to practical problems of spaceships, Tsander proposed in a speech before the Society's research committee on 15 July 1924 a twelve-point plan which included: testing small rockets powered by different fuels; testing rockets inserted into one another (i.e., step-rockets); constructing and testing folding and nonfolding aircraft models of various types, propelled by rockets and reciprocating engines, or both; testing high-acceleration effects in centrifugal machines (i.e., "g" forces); and constructing and testing liquid propellant engines or engines run by "solar energy." But even though a laboratory was eventually established, Tsander's ambitions came to naught.[25]

The Society for the Study of Interplanetary Communications was mainly a debating club. Besides Lapirov-Skobolo's persuasive presentation, Professor Valdimir P. Vetchinkin, a respected aerodynamicist from the Central Hydrodynamics Institute, also gave a well-received talk on space travel in the Polytechnic Museum on 31 October. Tsander himself was always visible, either in the audience or upon the podium. Undoubtedly the most heated of these debates occurred during not one but three days, 1 October, and 4-5 October. That unprecedented event took place in the large auditorium of the Physical Institute of the First University (in the building on Mokhovaya Street). Posters were printed up especially for the occasion. It all centered around a ludicrous and totally erroneous report that the American Goddard had sent a rocket to the moon on 5 August of the same year (1924). Apparently it was a much delayed spin-off of the sensational publicity aroused by Goddard's *Method of Reaching Extreme Altitudes,* first released to the general public in 1920, which sadly earned Goddard his newspaper cognomen "Moon professor." The same outrageous publicity elicited offers from scores of individuals earnestly seeking passage to the Moon and even to Mars on his rocket. The usually sedate *New York Times* joined in the chorus and published an open letter from Captain Claude R. Collins, a wartime pilot and president of the Aviator's Club of Pennsylvania, who declared himself ready to fly to the Moon or Mars if the professor would provide a $10,000 insurance policy on his life. A few days later the paper reported that a young lady from Kansas City, Missouri, was willing to accompany him.[26]

Now it was the Russians' turn. The Society for the Study of Interplanetary Communications might not have had its aspirant astronauts in the same outlandish manner, but they were the victims of over-zealous newspaper copyists nonetheless. So many people packed the hall on Mokhovya Street to hear the details of Goddard's supposed Moon rocket landing that the horse militia had to be called out to keep order and the conference had to be repeated twice more. Just how Konstantin Tsiolkovsky, Yakov Perelman, F. E. Dzerzhinskii, and other eminent honorary members of the Society regarded these proceedings we do not know. Tsiolkovsky communicated with Secretary Leiteisen from time to time but was for the moment probably spared the fiasco. Tsiolkovsky was then 67 and far away in Kaluga immersed in more serious and profound thoughts on the conquest of interplanetary space.[27]

The great debate may have speeded the death knell of the newborn Russian Society. Lack of finances and the unsettled state of the nation, then just out of the throes of civil war, also exacted their toll. Two later Soviet rocketry pioneers, B. V. Raushenbakh and Yu. V. Biryukov, sum up the demise of the Society for the Study of Interplanetary Communication when they say: "Quite a group of young fledglings in the scientific community during the early 1920s insisted on immediate space flights, but most of them failed to see the difficulties implicit in the proposal, and gave up when the task became hard. That was how the Society for the Study of Interplanetary Communications came to its end in 1924, to be followed in its fate by the Interplanetary Section of Inventors in 1927, and by other space-oriented circles and groups."[28]

In 1925, just a year after the premature collapse of OIMS, Academician D. A. Grave founded a new "space studies society in Kiev." In actuality, Dr. Dmitri Alexander Grave, whom we also meet later, was then not a member of the USSR Academy of Sciences but of the Academy of Sciences of the Ukraine to which he had belonged since 1919. Only later was he named an honorary member of the Academy in Moscow. Grave was a renowned mathematician by 1925 and had contributed much to problems of differential equations, cartographic projections, and other branches of applied mathematics. However, his impressive curriculum vitae in the *Bol'shaia Sovetskaia Entsiklopediia* (The Great Soviet Encyclopedia) fails to mention any details of his Ukranian astronautical group and Glushko only adds that among the individuals of its "research council" were "Academician" B. I. Sreznevsky, Y. O. Paton, K. K. Seminsky, and V. I. Shaposhnikov. Yevgeny Oskarovich Paton was the most famous of these men from the technical standpoint. The son of the Russian Consul in France in the latter half of the 19th century, Paton was also a member of the Ukranian Academy of Sciences. His speciality was bridge engineering and welding. The "Ukranian Astronautical Society," as we may call it, teamed up with the Kiev Association of Engineers and Technicians and held a space-flight exhibit that opened (presumably in Kiev) on 19 June 1925. It was hailed as a great success though no details are known. Somehow news of it even escaped Rynin who was a clearing house for any and all astronautical-rocketry developments in the USSR and elsewhere. The Ukranian Astronautical Society thus also quickly passed out of history.[29]

Rynin only says there was some "reanimation" to again unite the space travel idealists in the Soviet Union within not too many years after the collapse of OIMS in 1927. On 30 January of that year the Association of Inventors made the following appeal:

"The Interplanetary Section of the Association of Inventors calls your attention to an exhibition which will be held on 10 February 1927 at the Association of Inventors Building, 68 Tverskaya, Moscow. This is the world's first exhibition of models and mechanism of interplanetary vehicles constructed by inventors of different countries. The Association knows of your work on the problem of cosmic flights and believes you will not refuse to participate in our exhibition by submitting copies of manuscripts or published works in addition to sketches, models, diagrams and tables. Many inventors have already sent us material, among them the esteemed K. E. Tsiolkovsky, and from abroad we expect to hear soon from Robert Goddard of the United States, Esnault-Pelterie of France, Max Valier of Germany, Hermann Oberth of Rumania and [Ernest] Welsh of England. We would appreciate your material well in advance of the opening, but if for some reason this is not possible, please notify us."[30]

Mr. Welsh of England could hardly be included among the ranks of Tsiolkovsky, Goddard, Esnault-Pelterie, and Oberth. Welsh, a resident of North Ferriby, East Yorkshire, had neither devised a peaceable spaceship nor conceived a theoretical advance in space travel. Rather, he had invented a terrifying "death rocket" that threw out a shower of molten metal pellets against attacking air forces. The rocket, he claimed, could climb to a height of 8 km (five miles) and was successfully tested before British authorities at Hull in the summer of 1924. The British were impressed, as were the French and Americans. An American Chemical Warfare Service officer, Major Atkinson, was present for the trial and promised to transport one of the smaller rockets back to the United States via steamship, but difficulties were met in persuading the vessel's owners to forward such a dangerous cargo. Nothing further was heard of Welsh's "death rocket" though the Russians did manage to obtain a model of his supposed "rocketship" proposed in 1922 and propelled by melonite detonating in compressed air. This was proudly displayed at the 1927 exhibition in the Association of Inventors building on Tverskaya Street but apparently with no hint that it was a ground-to-air weapon rather than an interplanetary vehicle. Even so, Ernest Welsh was not representative of England's contributions towards spaceflight. A. M. Low, who afterwards became a force in the British Interplanetary Society, was probably Britain's leading advocate of interplanetary travel from the mid- to late 1920s but was strangely not consulted by the Russian Interplanetary Section of the Association of Inventors for their 1927 exhibit.[31]

Despite financial and "other difficulties," the First World Exhibition of Interplanetary Machines and Mechanisms was held between April and June of 1927. Its organizers were O. Kholoshchev, I. Belyaev, A. Suvorov, G. Polevoi, and one Pyatetskii. Ukranian Academician Dmitri Alexandr Grave bestowed not only his blessings from the world of academia, but submitted his own ideas on the possibility of trapping electrons and cosmic rays for space propulsion. "Many social circles have a skeptical attitude toward the subject of space research and the conquest of space," he wrote. "People think that they are associated with imaginary projects of space travel in the spirit of Jules Verne, Wells, Flammarion and other science fiction writers. A professional scientist cannot share this point of view. As long as five years ago, on the pages of the newspaper *Kommunist,* I pointed to the need of exploiting the electromagnetic energy of the Sun. The only practical approach to the utilization of the electromagnetic energy of the Sun was outlined by the Russian scientist Tsiolkovsky, who described in detail reactive devices or interplanetary vehicles as both timely and useful."[32]

N. A. Rynin's *Interplanetary Flight and Communication* contains a complete list of the displays at the exhibition. Among them were the Jules Verne-H. G. Wells corner; the Tsiolkovsky corner, including a bust of the pioneer; the Goddard corner; the corners devoted to Oberth and Valier, including one of Valier's rocket models and his publications; models of rockets invented by REP, Welsh, and the French writer Henri de Graffigny; the electron-propelled spaceship of the Austrian Franz Abdon Ulinski; a cross-section of G. Polevoi's rocket car and a scheme of his space station; A. Ya. Fedorov's atomic rocket ship showing its engine compartment and temperature regulator and overall view; G. Krein's corner featuring a vehicle propelled by electrical energy; Tsander's corner with a model of one of his rockets; a space suit; descriptions of radio wave and sunlight telegraphy by interplanetary ships to the earth; and various artists' concepts of rockets passing through star clusters, spiral nebulas, meteor streams, and cosmic radiations.

The reviews in Moscow were favorable. Elsewhere, The First World Exhibition of Interplanetary Machines and Mechanisms was largely unknown. No mention of it appears in the published papers of Goddard nor in the public writings of Oberth and Esnault-Pelterie. Max Valier's biographer alludes to a Russian newspaper account of Valier's invitation to the event but suggests that he was unable to attend because of nationalistic reasons. There is no indication that any foreigners attended at all. The publicity was thus confined to Russia. But the reporter, Salomeya G. Vortkin of the *Rabochaya Moskva,* was so elated at the prospect of interplanetary travel after seeing the exhibit that he volunteered his services. "I am going to accompany you on the first flight," he told the Interplanetary Section of the Association of Inventors. "I am quite serious about this. As soon as I heard what you had done, I tried in every way to make certain that you would take me with you." Setr, an artist at the Third Government Cinematographic Studio also reviewed the exhibit. "It would be desirable," he said, "that our [Soviet] inventors achieve the first landing on the Moon . . ." Thirty-two years later Setr's wish was fulfilled. On 14 September 1959 the 390.2 kg (858.44 lb) Luna-2 space probe reached the surface of the Moon, becoming the world's first flight to another celestial body.[33]

Austrian Groups

In 1926, in Austria, another dedicated group of individuals banded together to help fulfill the common aim of penetrating outer space.

This was no group of fledglings. Dr. Franz von Hoefft, the founder and chairman, was a 44-year old chemist of considerable experience who had obtained his doctorate at the University of Vienna in 1907. Before World War I he had simultaneously been a blast furnace engineer in Donawitz with the Vacuum Oil Company and a pretester at the Austrian Patent Office. After the war he was a tutor and private consultant. It was then that he also thrust himself fully into the theoretical aspects of interplanetary flight. Von Hoefft really began occupying himself with the question of space travel as early as 1891 and in the design of an "ether ship" from 1895. His fascination with the possibility of space-flight never waned, although he did not publish any of his ideas until the Oberth era. From then, his interest had been awakened anew and in 1926 led to the creation of the *Österreichische Gesellschaft für Hohenforschung* (Austrian Society for High Altitude Exploration).[34]

Baron Guido von Pirquet was named Secretary of the Society and also became one of the great names in astronautical theory. Born in his family castle at Hirschstetten near Vienna in 1882, he later qualified as a mechanical engineer at the technical high schools of Vienna and Gratz where he studied machine construction. In 1926 von Pirquet's technical competence was such that he was elected Vice President of the Technical Examinations Committee of the *Österreichischen Erfinderverbandes,* the Austrian Society of Inventors, an organization having close ties with the Patent Office. It was thus that he came into contact with von Hoefft.[35]

A "Rocket Committee" of the Society met irregularly at von Hoefft's home at Darwingasse

34 in Vienna and was also known as the Von Hoefft Committee, though a certain Engineer O. P. Fuchs was its real leader. Sometimes the group met in the Urania Observatory in Vienna. Among other members of the Committee were the chemist Dr. Korner; Professor Dr. Kirsch, a radiologist; Professor Wagner, a geophysicist; and Professor Karl Wolf of the College of Science and Technology in Vienna. "It just happened I went there too, once," recalled von Pirquet, and in a short time the Committee evolved into the Society of High-Altitude Exploration, or what may also be called the Austrian Rocket Society. Von Hoefft registered the new Society and was designated President and von Pirquet Secretary.[36]

The actual founding date of the *Österreich Gesellschaft für Hohenforschung,* eludes history. Von Hoefft seems to have been the first to propose a space travel organization in his country, from about March 1925. Max Valier, Western Europe's most zealous space travel promoter flatly rejected the idea. In a circular letter written on the 29th of that month to von Hoefft, Professor Wolf, Oberth, and Walter Hohmann, the city architect of Essen, Germany, whose book *The Attainability of Celestial Bodies,* had just been released, Valier suggested "a society or an association" was too premature. The principles should get to know each other first. Also, he concluded, "the less the outside world notices such an intellectual ring, the more efficient it is . . ."[37]

Valier's point of a trial acquaintanceship was well taken as the subject was not broached again until a year later. Valier, says his biographer, then agreed to the proposal; he again acted as more of a mediator-agent by forwarding von Hoefft's newest recommendation to Oberth and Hohmann. Von Hoefft wrote: "I must see in the space ship more than the only salvation [of mankind] but also the only justification of humanity and its culture." There were also "many first officers and staff officers" in the movement, he added, "but no field marshals,"[38]

Von Hoefft did not mean to imply he would assume dictatorial powers over any organization that might be formed, but that the space travel movement was fragmented and needed direction. Von Hoefft's "field marshal" ambitions seemed satisfied when the Austrian Society for High Altitude Exploration was finally formed by the Spring of 1926, but it was soon apparent that fragmentation still existed and clashes of ego were bound to occur, especially over the wide differences in approach to the space travel problem. Valier was not an organizer of the Society but was soon welcomed; Professor Oberth was not. On 22 June 1926 Valier wrote to Oberth of his progress: "Enclosed are my letters of 12 June and 22 June to the Society for High Altitude Exploration, with which I came into contact on 1 June, as you see with a certain success." To this, Oberth replied: "I thank you for having mentioned my name to the Society for High Altitude Exploration. I shall make an effort to satisfy the gentlemen . . . The Society for High Altitude Exploration has not invited me to join them. In any case, I have no desire to be the first person to write. If you write to the Society . . . please mention that in the meantime, many simplifications of my tests have occurred to me and that with approximately 9000 M [arks] I am capable of carrying out all the preliminary tests which are necessary for the construction of a rocket plane and of simple recording rockets . . ."[39]

What transpired thereafter is still largely hidden, though it is clear that Oberth was not accorded the full attention he deserved. By the end of that month the organization may have undergone a reorganization, as it was sometimes called the *Gesellschaft für Weltraumforschung,* or Society for Space Research.[40]

In any case, Valier suggested a conciliation of differences between Oberth and the Society. On 27 December he told the Professor, " . . . it may be of little use to you, but it would prejudice me a great deal if I were to put my series of pictures at your disposal. Since you have already found a support in the Society of Vienna, I think it would be best if you make a settlement with this Society about your lectures . . . Moreover, if you have read somewhere in newspapers that I want to have myself shot to the Moon with a rocket you should know that this announcement in no way comes from me but my opponents, who are trying to make my actual plans riduculous."[41]

In his reply of January 1927, Oberth brings the picture into sharper focus. "As far as your remarks on the Society for the Exploration of Space are concerned," he wrote, "I must say that you regard the situation, especially my status in the Society, from an incorrect point of view. [Von] Hoefft had called in various people who have a well-known name but who, in addition, do not have the faintest idea of the cause itself, to bring the Society more prestige . . . Therefore I must come to Vienna of my own accord and press the Society hard if I want to bring them to their senses and to do away with a few superfluous people. [Von] Hoefft did not dare to submit to the

Society a letter which I wrote to them and in which I pointed out why it lay in our mutual interest to make the journey in Vienna possible for me . . . "[42]

Professor Oberth did not go to Vienna. Otherwise the aftermath of this discord has been unrecorded. In fact, little else has been published of the fate of the Austrian astronautical community thereafter. Not only were its leading members, von Hoefft and von Pirquet, more directly associated with German Rocket Society affairs when that organization was founded in July 1927, but the German Rocket Society overshadowed all other groups or activities in non-Russian Europe from there on. At least one exhibit of the Austrian Society for High Altitude Exploration is known. It is reported they had a stand at the Aeronautic Exposition of the International Show at Vienna, from 11—17 March 1928. An attempt at experimentation was also undertaken. It concerns another great Austrian pioneer, Eugen Sänger. Guido von Pirquet later wrote about it: "In 1927, [von] Hoefft had the idea to have a rocket model tested in the wind tunnel of the Institute of Aerodynamics at the Technical University of Vienna. Based on the concepts of [von] Hoefft, I built the test model. While the test results were satisfactory, they did not find any immediate technical application. But we learned at that time that a young assistant of the Institute was a great rocket enthusiast. Thus, for the first time, I heard of Eugen Sänger. Somewhat later I learned that Sänger was looking for a place to test rockets. As I owned a vacant field near Vienna, 1 km in length and 140 m wide, which I considered suitable for such tests, I contacted him and he came to see me and my wife in Hirchstetten and we met personally for the first time. However, the tests were not made on my property after all and I also did not discuss with Sänger the possibility of testing my nozzle configurations."

Sänger applied for membership in the Society on 27 March 1928 and offered his assistance for von Hoefft's preliminary experiments at the Institute for Aerodynamics. Nothing apparently came of this proposal and Sänger went to conduct his work either privately or under governmental or Technical University sponsorship.[43]

In March of 1931 the Austrian Society was superseded by the Österreichische Gesellschaft für Raketentechnik (Austrian Society for Rocket Technology). The new organization was founded by von Pirquet and Rudolf Zwerina, a man apparently well known in Austrian aeronautical circles. Von Pirquet delivered his inaugural speech in Vienna on 16 April 1931, but otherwise little was heard from the Austrian quarter. Brügel, in his Männer der Rakete (1933), says that the inaugural meeting took place in the Pavillion of the Österreichischen Erfindersverbandes (Austrian Inventors Society). In fact, the rocket group was part of the Inventors organization and shared its headquarters in the same building at Wein I (Vienna I), Postgasse 7. When Brügel wrote, von Hoefft was President; the two Vice-Presidents were Friedrich Krauss, who was also the President of the Inventors Society, and von Pirquet. Zwerina was Secretary. Perhaps these were the original officers of the Austrian Society for Rocket Technology. In any event, the Great Depression had hit Austria as it had other countries world-wide so that few could afford to support the rocket work. The Inventors Society also lacked facilities. In a letter from Willy Ley, the German rocket pioneer, to G. Edward Pendray, his contemporary, dated 2 November 1931, Ley remarked in his quaint English that: "From Baron Guido von Pirquet I have heard, that the Austrian Society doesn't come to work, because they can't get money, it is in Vienna like in Berlin but we [the German Rocket Society] have our Raketenflugplatz and they have nothing."

Von Hoefft's name had already been etched into the foundations of the movement, nonetheless, and his place in history assured. He had not only instituted a Society and set the stage for others, but also extensively contributed to the literature in his own right. Most notable were his lengthy articles appearing in the German Rocket Society's journal Die Rakete in which he worked out an entire program of rocketry research from a simple balloon-borne liquid oxygen-alcohol "recording rocket" designed to fly upwards of 100 km (62 miles)—the RHI, or Rakete-Hoefft I—to a Moon, Mars, or Venus-bound RHVII rocket. As early as 1928 he also spoke of mail transportation rockets and rockets for automatically taking pictures of the Moon and other heavenly bodies. It was unfortunate that also within the pages of Die Rakete is found a heated and even acrimonious "technical dispute" he had with Oberth. The editor, Johannes Winkler, felt compelled to arbitrate. "I call your attention," he wrote, "to the fact that in the future I shall refuse contributions which do not preserve an academic tone, no matter who is the author."

As for Guido von Pirquet, he continued to write prolifically on astronautics and rocketry until his death in 1966 at the age of 86. In fact, von Pirquet's name appeared in almost all the journals of the later astronautical societies. This included not only the German Rocket Society's Die Rakete but also the journals of the successor organizations, as well as American and British Society publications. The international careers of such pioneers indeed attest to the interaction of the societies and the value of their literature. Yet apart from von Pirquet's articles, virtually nothing was published in magazines such as *Das Neue Fahrzeug* about the Austrian group itself in the mid-1930s. Its final days were as much of a mystery as its beginnings.[44]

The final chapter of the Austrian rocket group was written when the country itself was dissolved in the German take-over, the *Anschluss* of 1938. Whatever organization remained must have ceased to exist altogether. Perhaps by its early example, and certainly by the direct intervention of Max Valier, a newer, larger, and more durable group came into being not long after von Hoefft started the *Österreichische Gesellschaft für Hohenforschung.* This became popularly known as the German Rocket Society.[45]

IV

The VfR

The Breslau Years

The Society for Space Ship Travel, or *Verien für Raumschiffahrt*—the VfR, and afterwards more popularly called the German Rocket Society—was born on July 5, 1927. It met in the parlor of the *Goldnen Zepter* (Golden Scepter) tavern on Schmeidebrucke 22 in the German industrial town of Breslau (now Wroclaw, Poland). At least nine men and one woman were present at the opening meeting. One of the official founders, Willy Ley, was absent. Present were Max Valier; an engineer from Berlin, Johannes Winkler, an engineer then working as a church administrator; Georg Lau, member, District Board of Works; Theodor Fuhrmann, a clergyman; Alfons Jakubowicz, candidate for a chemical engineering degree; Miss Hedwig Bernhard; Gerhard Guckel; Herbert Fuchs, a pastor from the nearby town of Nestau bei Suhlendorf; Walter Neubert of Munich, probably a friend of Valier; and an engineer from Berlin, H. Sauer.[46]

The late Willy Ley wrote about the events that precipitated the formation of the Society. Reminiscing in a semi-autobiographical article he wrote in the magazine *Space World* for June 1961, Ley recalled: "In 1927 (just before our meeting), I received a letter from Max Valier, who was lecturing a great deal. He suggested that a club be organized to raise money and finance rocket experiments for Oberth. Such a club would need a legal charter and Valier asked that I contact a man in Breslau by the name of Johannes Winkler who would make the necessary court applications. Winkler complied . . . I was not personally present at the first meeting and was, therefore, not listed in the charter. But I had been active in its formation and, of its original founders, I am the only one now [1961] alive. Professor Oberth, Dr. Hohmann, and Dr. von Braun—all alive today—joined the Society later and were not founders."[47]

Earlier, in the spring of 1927, apparently on April 29, Valier delivered a lecture before the *Wissenschaftlichen Gesellschaftlichen für Luftfahrt* (Scientific Society of Aeronautics) in which one of the audience, Stephen von Prodczinsky, a former naval officer and aviator and then (1927) departmental head of the German Test Plant for Aeronautics, asked leave to speak. He proposed that all those concerned with rocket problems should "team up in order to work together to the same end, exchanging their ideas, in order to avoid unnecessary wasting of funds." Valier only replied that his own work had thus far been conducted out of pocket and he also expressed a certain self-satisfaction that he had been able to manage by himself thus far. But the idea of a group quickly took hold and he soon contacted Winkler.[48]

How he knew Winkler is unknown. Valier traveled widely on his lecture circuit and certainly knew of others with the same interests. He was so busy lecturing that when the historical July 5 meeting in the Golden Scepter was held, he declined the Presidency on the new-born organization because of his speaking commitments.[49]

The Breslau Court was taken aback by Winkler's request for registration. Not only did the group barely meet the minimum membership requirement of seven to carry the letters e.V. (*eingetragener Verein,* or registered society), but the Court at first refused to admit the full name, *Verein für Raumschiffahrt, e.V.* because "the aims of the proposed association will not be apparent to the public, since the word space-travel does not exist in the German language." The Court relented on the grounds that the definition of this phrase be included in the Articles of Association and that new inventions require new words.[50]

The charter worked out at the Golden Scepter was as expansive in its goals as the idea of space travel itself. "The purpose of the union," they wrote, "will be that out of small projects, large spacecraft can be developed which themselves can be ultimately developed by their pilots and sent to the stars." Typical of the Society's early slogans was "Help create the spaceship!" They also optimistically suggested that "the work of Valier with light aircraft" might be the path towards these spaceships and also that Herr [Walter] Neubert, Munich, clearly states that he will also try to attempt this feat with a pure rocket apparatus." Valier later constructed and rode rocket-propelled cars, sleds, and rail cars, but not planes. Nothing further was heard of Neubert's project. It was also ambitiously stated in the first minutes that "through careful management" 200,000 Reichmarks might be raised by dues and donations to finance the spacecraft.[51]

The protocols of the charter itself were more mundane, dealing with the name of the new organization; the official grant of registration or incorporation; the use of the calendar year for conducting business; reporting procedures to the governing committee; annual dues; the organs of the VfR administration; the governing committee and the membership; duties of the President; the VfR as an essentially money-making venture; authorization of the publication of

Die Rakete; the nature and duties of the executive or governing committee; and the planned change of the executive committee that fall.

The first committee consisted of Valier, Winkler, Fuhrmann, Jakubowicz, Sauer, and Neubert. Theodor Fuhrmann became Treasurer. The Presidency and editorship of *Die Rakete* was reserved for Winkler. Winkler ran *Die Rakete* throughout its three year existence and remained as President of the VfR until Oberth assumed the seat in the Fall of 1928.[52]

Rector Fuhrmann was also responsible for enrolling new members. He must have done his job well because within a year the Society boasted of 500 members. By September 1929 the number had expanded to 870 and rose to more than 1,000 soon afterwards. The 15 October 1927 issue of *Die Rakete* reported that about 20% of the membership were engineers.[53]

Donations of both money and supplies came in, but there was never enough for the VfR. Donation lists were regularly printed in the journal and an examination of the names of the donors and their residences reveal the truly international nature of the group. They came from Poland, Czechoslovakia, Russia, France, Denmark, Spain, South America, and even from South-west Africa where a ten-man astronautical society had been set up there within the German community in 1927 by J. Konetzny. Almost all of the donor's names are Germanic.

More important to the life of the VfR was the donor Hugo A. Hückel of Neutischein, who conveniently owned an aluminum factory and also made hats. It was this same Hückel who financed the first large rockets constructed by Winkler, known for that reason as the HW Series. Hückel generously contributed to the VfR throughout. Also on these lists are all the great names in rocketry and astronautics of the period, including Esnault-Pelterie, Perelman, Rynin, Hohmann, Oberth, Hermann Potočnik (pseudonym for Hermann Noordung, author of the first book on space stations), von Hoefft, Sänger, and Sander (constructor of von Opel's powder rockets for his cars). The name of von Braun also appears. Winkler and his family also gave money out of their own pockets as did Ley and Valier.

Incentives were given to members to enroll new members. Thus, those who signed up three people were given an autographed picture of Max Valier; those who signed up five persons were entitled to an autographed offprint of Valier's lecture, *Die Fahrt ins All* (The Trip into Everything) [i.e., space] or an autographed copy of Willy Ley's *Die Fahrt ins Weltall* (The Trip into the Universe). For those who acquired ten new members, gifts of Valier's *Der Vorstoss in den Welten-raum* (Advance into Space), 4th edition, 1928, autographed, were given.[54]

An even more ambitious plan was concocted whereby, after reaching a highly optimistic membership level of 10,000, the person who enlisted the largest number of new members was to be rewarded with a 2,000 Reichmark prize, (later increased to 5,000), the second with 1,000 RM, the third with 500 RM, and so on.[55]

The problem of raising capital became more acute as the Depression set in. So did the ingenuity of the VfR. A sort of prospectus dated 6 October 1931, was sent to the President of Noyes Buick Sales in Boston—and perhaps other firms—outlining their "rocket shows", which could be obtained for certain set fees. The basic show, excluding the cost of insurance and freight, came to $500.00. This included: 1) a high-altitude 15 meter (49 ft.) long rocket; 2) the same, with parachute; 3) "the first rocket for liquid propellants, built in 1928"; 4) a 1928 (i.e., von Opel) rocket car; 5) a portable "proofing stand for rocket motors"; 6) a model of a spaceship "able to reach the moon or another planet of our solar system"; and "our 'museum', consisting of all our rocket-motors and another apparatus from the beginning of our work." There is no evidence that the President of Noyes Buick Sales, nor any other American ever took up these offers.

Ernst von Khoun formerly with the *Bayerische Rundfunk* (Bavarian Radio), says that additional publicity, if not revenue, for the VfR may have been gained through the good offices of the science fiction writer Otto Willi Gail. In the early 30s Gail was a nationally known "science-nature" commentator over the *Deutsche Stunde in Bayern* (German Hour in Bavaria) radio show. While insisting on factual rather than "fictional" presentations, Gail confidently maintained that the dreams of Valier and the VfR would become a reality. For this reason he seems to have reported their progress from time to time. A few documentary movies were also made for public consumption, one being a *Ufa Tonwock* (Talking Week show) newsreel of later VfR experiments. Still another publicity and fund raising scheme came in the summer of 1931, when Willy Ley wrote to his American friend, G. Edward Pendray, and reported in his inimitable English that: "Now we have also pins for the members of the *Verein für Raumschiffahrt,* a silver space-ship on a black ground with the name of our Society. If you like to get one (or two, for Mrs. Pendray too) please, write me. Price Mark 1, 50 the pin.''[56]

The VfR's international and scientific reputation was especially enhanced with the memberships of Oberth, Rynin, Perelman, Esnault-Pelterie, Hohmann, and von Pirquet. In an announcement in *Die Rakete* for 15 November 1927, both Professor Oberth and Walter Hohmann, under the regulations of the VfR, were formally declared members of the directorship.

The first issues of *Die Rakete* also began to run leading articles by guest authors in a fairly diversified and informal manner. The unsigned ones were probably by Winkler and Ley, the co-editor. The opening article was ''The Flight to the Moon—Its Astronomical and Technical Basis.'' Other pieces included biographical sketches of prominent pioneers such as the Austrian theorizer of electron propulsion, Franz Abdon Ulinski, and the influential German science fiction author Otto Willi Gail.

Die Rakete however had a predecessor, a magazine printed six months before the VfR came into being, the *Deutsche Jugend-Zeitung* (German Youth Journal). The lead article was, interestingly, *''Der Flug zum Monde, seine astronmischen und technischen Grundlagen''* (''The Flight to the Moon, its Astronomical Basis''). The fourth issue contained a subtitle change: ''Vereinigt mit der Zeitschrift '*Die Rakete*' '' (''Combined with the journal '*Die Rakete*' ''). This journal ceased publication shortly thereafter. When the VfR came into being a new *Die Rakete* appeared, along with a supplement containing articles from the original magazine.

From the beginning the VfR was intent upon experimenting. In time their initially grandiose plans for space rockets were cut back, but the Society did attract a talented and dedicated coterie of technicians and theorists who laid the groundwork for future space vehicles. Dating the first VfR experiments is difficult because Winkler did much of his early work independently while still associated with the Society. He also had the assistance of fellow members. Alexander Scherschevsky, a Russian aeronautical student and writer residing in Germany and member of the Society, wrote that the first VfR experiments were made 23 November 1927 at Heidelwilen bei Obernigk in Silesia with small skyrocket-propellant model biplanes weighing 200 grams. However, the *Breslauer Modell-und Segelflugvereins* (The Breslau Model and Sailplane Society) was actually responsible, but supported with VfR procurement of the rockets. Hermann Oberth also undertook some experimental work, also on a private basis, for the Ufa movie company by whom he had been hired to construct a flying rocket for the publicity of the 1929 motion picture *Frau im Mond* (Woman in the Moon). The first recognized VfR rocketry work began in 1930, near Berlin, as an offshoot of Oberth's project.[57]

By that time, Johannes Winkler had severed his connection with the VfR. He was conducting an independent program of making solid propellant thrust curves at the machine laboratory of the *Technische Hochschule* of Breslau (the results of which were detailed in *Die Rakete* for 15 January 1928). With the financial backing of Hugo A. Hückel, Winkler succeeded in sending aloft what was hailed as the first liquid-fuel rocket in Europe on 21 February 1931 at Dessau. Significantly, Willy Ley recalled that, ''The first European liquid-fuel rocket (since Goddard had not then published his second Smithsonian report [of 1936] we naturally took it to be the first liquid-fuel rocket [flight] anywhere . . .''[58]

Also unknown at the time, pyrotechnist Friedrich Sander appears to have flown his own liquid-fuel rocket—in secrecy—as early as 10 April 1929. Winkler's work ultimately led the Junkers airplane company to sign a contract for the application of liquid-fuel rockets for the assisted take-off of aircraft such as their Bremen type seaplanes.

Winkler resigned his presidency of the VfR by the Fall of 1929, ceding the chair to Oberth, and also ceased editorship of *Die Rakete*. The magazine ceased publication. The time had come anyway for the VfR to choose between financing a journal or an experimental program. P. E. Cleator, editorializing in the *Journal of the British Interplanetary Society* for April 1934, summed up the situation: ''In the year 1929, the old German rocket society, the *Vereins fur* [sic] *Raumschiffahrt E. V.,* [sic] ceased the publication of their journal, *Die Rakete*. The immediate result was the loss of over six hundred members! . . . Now the new experimental programme was all very well for those members who happened to live in Berlin, for they could take part in, or witness, the experiments. But not so the majority of the members who were scattered throughout the country. With the loss of the journal, they were deprived of their only real link with the Society.''

Coeval with the extinction of *Die Rakete,* Valier had gone off on his own tangents. In collaboration with various individuals, notably the publicity-seeking automobile magnate Fritz von Opel; the Wessermunde pyrotechnist, Friedrich Sander; and finally, with Dr. Paul Heylandt, the liquid-oxygen manufacturer, Valier embarked upon a dizzying, misunderstood career of riding rocket sleds and rocket cars. He too progressed towards liquids, the explosion of the rocket engine for the liquid-fuel *Rak 7* automobile on 17 May 1930 killing him.[59]

Meanwhile, in 1928, a new figure had joined the ranks of the VfR and was to change the entire course of the organization. Rudolf Nebel approached the Society, or at least one of its members, Professor Oberth, through a newspaper advertisement. Oberth, though a brilliant theorist, was neither an organizer nor a mechanic. Faced with the heavy responsibility of serving as a technical advisor for Fritz Lang's movie, *Frau im Mond,* as well as building and flying a real liquid-fuel rocket for the film's publicity, Oberth cast about for assistants and placed an advertisement in several daily newspapers. A 34-year-old man from Bavaria was one of the applicants and introduced himself: "Name is Rudolf Nebel, engineer with diploma, member of the oldest Bavarian student corps, World War combat pilot with rank of lieutenant and 11 enemy planes to my credit." Nebel was hired at once and apparently was never interviewed further as to his real qualifications. As it turned out, he was more of a master manipulator and operator than an engineer. His ego and aggressiveness more than made up for his slight stature. Willy Ley afterwards revealed that: "Nebel himself told me later, without regarding it as a personal secret, that he had been graduated in a hurry during the war because he had volunteered for the air arm, and that after the war he had never worked as a designing engineer but for some time as a salesman of mechanical kitchen gadgets instead. Since jobs were almost impossible to find, all this was probably not his fault—but he was not the man Oberth needed."[60]

Ley and others generally fail to mention that Nebel claimed also to have actually experimented with rockets, but with questionable success, as early as 1916. In his 1972 autobiography, *Narren von Tegel* (Fools from Tegel), he claimed that while recovering from wounds suffered from the crash of his Fokker monoplane after being hit with enemy fire near Cambrai, in the Somme, he first thought of rockets on planes. Upon his release from the field hospital he went to the nearest engineering supply depot and drew out ten 1-meter (3 ft.) long silver-grey signal rockets, along with "spring heads," cables, and other materials. He attached four open tubes to the undersides of the wings of his new plane, an Albatross D III, two tubes on each side. Then, he says: "Everyone was tensely waiting to see what would happen when, 4,000 meters [13,124 ft.] up, we encountered an enemy squadron of 25 planes. I pressed the button; an immense trail of powder smoke passed through the center of the enemy squadron [of English biplanes]. One plane immediately dropped its nose and went into a dive, landing on the nearest meadow." The English pilot was captured. Eight days later Nebel's rockets again downed some airmen. "With the second hit," he says, "I succeeded in shooting away an enemy propeller. Only with the third shot did I get into trouble; I shot myself down. When I pressed the button, the primitive hand-made rockets exploded before they had left my airplane. The aircraft caught fire—and there weren't any parachutes at the time. I plummeted toward the earth in my burning machine. But then I had an improbable stroke of luck. Right next to where the burning aircraft struck the ground, two privates were working on line construction and they got me out of the burning aircraft in time."

In his *Die Narren von Tegel,* Nebel also claimed that Hermann Goering, then a lieutenant in his unit (*Jagdstaffel* 5, or Fighter Squadron 5), heard of the new weapon and gave it a name: *Nebelwerfer* (Fog-thrower). Goering, the future chief of the German Air Force, did serve in *Jagdstaffel* 5 during World War I, but only for a matter of weeks, in the Fall of 1915. He was shot down by British Sopwiths and was in the hospital for several months. He did not return to the front until the summer of 1916, but as a pilot and latter commander of *Jagdstaffel* 26. Further, the Nebelwerfer of World War II had no connection at all with Rudolf Nebel. Nebel would have us believe otherwise, that his World War I air-to-air rockets were converted to ground-to-ground use in 1941; actually the *Nebelwerfer* was originally designed as a smokelaying mortar. This alone was responsible for its name.[61]

As for additional details of Nebel's career prior to joining Oberth and the VfR, he says that he managed to attend the *Technische Hochschule* (Technical High School) in Munich before the war, studying machine building. The war interrupted these studies. Following the war he returned to school and received his engineering degree in 1919, then worked briefly with the

Nürnberg Construction Bureau of the Siemens Company. By the end of 1920 he was a "senior engineer" with the German-Swedish SKF-Norma firm, manufacturers of ballbearings in Nürnberg. Here, he says, he made "good money" and learned the arts of selling and publicity on the "American style," selling roller and ballbearings. Nebel, meanwhile, did not lose his interest in rockets. In 1923 he invested in a partnership in a small fireworks factory at Pulsnitz, in Saxony. He says he could thus continue to experiment with powder rockets, though an explosion ruined that endeavor. He then went to Berlin, where he worked for two years in his friend's galvanic battery factory. In 1927 he was again employed by the Siemens firm selling burglar alarms. Following this, he made his real entry into astronautics, or at least liquid fuel rocketry, his introduction to Hermann Oberth. Then, he says, he became a member of the Ufa motion picture company staff and earned 600 Marks per month.[62]

The late Rudolf Nebel was Oberth's first hired assistant, while Scherschevsky, the Russian aviation student-writer, was the second. Nebel proposed building a small rocket with a 1.9 liter (half gallon) propellant capacity. Lang's movie company, Ufa, desired a larger projectile, at least 13.7 meters (45 feet) long. A compromise was settled on a 2.1 meter (seven foot) rocket holding 7.6 liters (two gallons) of propellant. Oberth, the high strung theorist, Nebel, the non-mechanic, and Scherschevsky, whom Oberth characterized as "the second laziest man I ever met," thus set about building their 15.5 km (25 mile) instrument-carrying high altitude rocket within three months. Running over schedule and winding up dismissing Scherschevsky in exasperation, Oberth barely completed a rocket and even that was considerably inferior in design much less capabilities. The episode, Oberth afterwards admitted, had been "disgraceful." "First, I was not a trained mechanic . . . Second, my nerves were almost shattered by an explosion in the Fall of 1929 . . . as a consequence of my tension and taut nerves I had committed several grand blunders, especially in treating people."[63]

Frau im Mond premiered 15 October 1929, at the Ufa-Palast am Zoo in Berlin, but the rocket was never launched. Out of desperation to meet his deadline, Oberth had opted for a cruder model for demonstration purposes only. It was a hybrid system utilizing solid carbon sticks immersed in liquid oxygen. The sticks burned from the top and produced a poorly conceived (stability-wise) "nose drive" propulsion of engine in the head. Oberth failed to find the right carbon-rich substance and the rocket never flew. Despite this setback, Oberth and Nebel had gained valuable experience through trial and error, and amassed a collection of tools and equipment, including an iron launching stand. Through the intercession of the ubiquitous Willy Ley, this gear was turned over to the Society. Nebel also joined the VfR through Ley and became its secretary. Ley actually introduced Nebel to the VfR's Berlin "representative", the patent attorney, Diploma-Engineer Erich Wurm.

A notice in Die Rakete for 15 August 1929 had announced that Wurm's office was available for use by the Society even when he was on vacation. From this beginning, the VfR began its Berlin phase and its own experimentation.[64]

Events moved quickly after 1930. Nebel and Oberth continued their work with the Ufa-bought equipment but now as a VfR function. First, funds were needed. In their hunt for likely sponsors, they came across the Chemische-Technische Reichsanstalt (the Reich Institute for Chemistry and Technology), which was somewhat equivalent to the U.S. Bureau of Standards. While no funds were forthcoming, the director, physicist Dr. Franz Hermann Karl Ritter, offered to test a VfR liquid fuel rocket. If it performed well he would officially register it which would go a long way in promoting the aims of the VfR to other institutions.

Oberth and Nebel were now joined by three new members, Rolf Engel, Klaus Riedel, and Wernher von Braun. Von Braun was then an 18-year old apprentice at the Borsig engineering works that specialized in railroads. He also attended school part-time at the Charlottenburg Technische Hochschule. Together, this team redesigned and reworked Oberth's Kegelduse ("cone-jet") motor, so-named because of the configuration of its steel and copper combustion chamber. Despite terribly inclement weather the test was run and Dr. Ritter's official affidavit affirmed that before witnesses the Kegelduse "had performed without mishap on 23 July 1930, for 90 seconds, consuming 6 kilograms [13.2 lbs.] of liquid oxygen and 1 kilogram [2.2 lbs.] of gasoline, and delivering a constant thrust of about 7 kilograms [15.4 lbs.]"

While not usually recognized as such, this may have been the VfR's first liquid rocket; it certainly was the first officially certified rocket anywhere. Shortly after, Professor Oberth returned to Medias, Rumania, to resume teaching.[65]

Prior to leaving, Oberth asked the respected architect and astronautical pioneer Walter Hohmann to assume the Presidency. Hohmann, who had written one of the first classic mathematical treatises on spaceflight, in 1925, declined due to work commitments. The VfR presidency thus devolved on Major Hans-Wolf von Dickhuth-Harrach, a retired Army officer who was much less well-known in space travel circles at the time but who was an ardent advocate and had written some popular space travel articles.[66]

The Berlin Years—Raketenflugplatz

In 1930, the VfR permanently moved its headquarters from Breslau to Berlin. The Society formally made its debut in the German capital by way of a public lecture in the auditorium of the General Post Office on 11 April 1930. Adorning the hall was an Oberth rocket dangling from the ceiling by parachute. Winkler delivered the principal speech. Two other founding fathers of the VfR, Valier and Ley, were also present, as was the eccentric and cantankerous German inventor, 74-year-old Hermann Ganswindt, whose remarkable 1891 proposal of a spaceship propelled by dynamite cartridges has already been described. He was, up to that time, perhaps more well-known for his horseless carriages, cycles, fire engines, and his progeny of 23 children. Also present were Erich Wurm, the original Berlin "representative" of the VfR, scientists and industrialists, Wernher von Braun, and of course, the irrepressible Rudolf Nebel.

The VfR also benefitted from the publicity of a 14-day public showing of their equipment and concepts coinciding with "Aviation Week," 25—31 May 1930, at the Potsdamer Platz and also in the basement of adjacent Wertheim department store. In the exhibit at the Wertheim—described as Berlin's largest variety store—were shown the Oberth rocket ready for launching in its tower; the rocket with its parachute open; performance diagrams showing the efficiency of rockets; a "rocket wheel" capable of revolving at 39,000 rpm (i.e., a dynamometer for testing rockets); gears and motors; photographs and books on astronautics; and a display of the use of wireless telegraphy reporting the whereabouts of the rocket at every minute. Outside of the privately built *Kegelduse* motor, which was turned over to the Society, no VfR rocket had really been built.[67]

During the period from June until September 1930, Klaus Riedel, Rudolf Nebel, and Kurt Heinisch were already constructing and testing the VfR's first *"Mirak"* or *"Minimum Rocket"* on a farm owned by Riedel's grandparents at Bernstadt, Saxony. Nebel, in his *Die Narren von Tegel*, says the three of them had been invited to spend the holidays in Bernstadt by Riedel's grandmother. Nebel's silver-grey Buick served as the transport. At the farm, Riedel's uncle also assisted in providing use of his own workshop. Rudolf Nebel provides us with the only known eye-witness account of the VfR's first rocketry work: "For our experiments we needed in Bernstadt liquid oxygen that I picked up in Gerlitz with the car. The small *Mirak* did not need much, and a container of the fuel lasted almost a week for us. It took quite a while until we fathomed the mysteries of automatic ignition. First, the propellant pressure must climb to ten atmospheres, then the fuel valve must be opened. The gas was driven with the help of the carbon dioxide cartridge [used to pressurize selzer water bottles] and then ignited. Finally, oxygen was added. With this sequence [operation] was obtained. Our protective room was a great straw pile behind which we went into cover when it was time. Through a large mirror the rocket could be observed from this place. With a so-called barograph the thrust of the *Mirak* was recorded, and we could read off the values for the pressure and time. In the first experiment in the field there was a thrust of only 400 grams [14 ozs]. Later, we raised it to 2 kgs [4.4 lbs] and after that to 3.5 kgs [7 lbs]. With this thrust, the rockets would have flown if we let them go [lossened them from their restraints used in the static tests]. But we wanted first to have some practice in burn tests. On 7 September [1930] we then prepared for the first take-off of *Mirak* I. The mayor of Bernstadt was invited and it was to be a great event . . . 7 September marked the end of [*Mirak* I]. The rocket, which should actually have flown for the first time on this day, exploded totally after ignition. Reidel and I had already recognized before in theoretical calculations the weaknesses of the one-liter *Mirak* I and now decided not to rebuild it. The experiments in Bernstadt had amply demonstrated that we must have a *Flugplatz* (flying place), and several work facilities and living quarters. My dreams of a rocket flying field must be realized." Nebel also says that Riedel "was always sending a few lines to Berlin about our work which Willy Ley then published in the Society news-letter." These have not been found. In any case, Ley's accounts published in his various books are very brief.

"These short reports nevertheless," adds Nebel, "had their effect. Several interested parties approached Wurm and donated towards 'liquid rocket' construction. One of the open-minded patrons was Hugo H. Hückel. He sent first 250 Marks, then promised to support our work monthly with 500 Marks. The conditions were that this money be used exclusively for experimentation. We naturally were very pleased about this *fortune* which then came in punctually every month."[68]

Nebel now determined to find a more suitable launching range, preferably closer to Berlin. Nebel and Riedel found an abandoned army garrison of 1.2 km² (300 acres) in the northern Berlin suburb of Reinickendorf. It was remote enough for safety's sake and possessed useful concrete bunkers where ammunition had been stored during the war. The Berlin municipality was persuaded by Nebel to lease the property to the Society for the nominal sum of 10 Reichmarks ($4) a year.

"Thus," Nebel afterwards proudly proclaimed, "on 27 September 1930, I established the first launch site in the world in Berlin-Reinickendorf." Klaus Riedel should rightly have been claimed as the co-founder, a fact that Nebel never properly acknowledged. It was hardly the first launch site but nonetheless was grandly dubbed the *Raketenflugplatz* (Rocket Flying Place). As the aggressive founder of *Raketenflugplatz* and the Secretary of the VfR, Nebel became the *de facto* President though that position was officially held by von Dickhuth-Harrach until 1933. Nebel and his impecunious friends thereupon devoted much ingenuity to scheming for free tools, aluminum and magnesium rods, welding equipment, paint, pipes, screws, a drill press, sheet aluminum, two lathes, benches, and a smithy. In his turgid autobiography, *Die Narren von Tegel,* Nebel himself delights in relating how he cut corners. It started from the first day. The morning after he, Riedel, and Heinisch moved into the vacant and drafty *Raketenflugplatz,* Nebel drove to the State Railway Office in Berlin and, through several trips, hauled back free lumber from abandoned railway cars. This material served to patch up decayed walls. With woodboard from the same source he covered the "ice-cold floors" and also found two old cannon stoves to warm up the place for the coming winter. A typewriter was next procured from a pawn shop. With this machine, hundreds of letters were sent to various companies and Governmental organizations seeking help. Some times Nebel's appeals were a bit too extravagant. The work of the *Raketenflugplatz*, he sometimes said, was "in the best interest of national defense and this is also the opinion of the *Reichsanstalt.*"

In 1931 Nebel also persuaded the tax office in Berlin to sign an agreement permitting the VfR to purchase gasoline duty-free, at 6 pfennigs per liter instead of 16. "Naturally," Nebel boasts, "we took advantage of this gasoline also for our [own] cars. Besides the Buick [Nebel's private vehicle] our personnel meanwhile used a motorbike with side-car and an NSU vehicle." The resourceful Nebel also persuaded various firms to donate free light and power, liquid oxygen, duralumin, and other items. Lufthansa also provided a gratis plane ticket, presumably in connection with a rocket or space exhibit. From von Braun, one of the non-paid helping hands in subsequent experiments, another view of Nebel's "acquisitional aptitude" is afforded. Nebel, he said, "once talked a director of Siemens Halske A. G. out of a goodly quantity of welding wire by vividly picturing the immediacy of space travel. Our own use for such wire was extremely small, but Nebel offered it to a welding shop in exchange for the labor of a skilled welder—which we badly needed. Our labor force cost nothing by reason of the then prevailing general unemployment. Many a draughtman [sic], electrician, sheet worker, and mechanic was only too happy to take up residence rent-free in one of our buildings and to maintain his skill at his trade. Soon there were some fifteen craftsmen living in our refurbished buildings and working eagerly on the tasks we set them."

Some of these individuals thus lived rent-free, spent their days at *Raketenflugplatz,* and paid for their food through welfare checks or got 15 pfennig meals at the Siemens welfare kitchen. The craftsmen remembered by von Braun were evidently the 15 personnel (including two women secretaries) Nebel obtained from the Labor Ministry's *Freiwilligen Arbeitdienst* (Voluntary Workers Service). These volunteers were subsidized by a small Government stipend if their otherwise free services were requested.[69]

Herbert Schaefer, who joined the VfR in the Spring of 1932, also says that: "We engineers received our meager pay from two sources: a) from the *Ingenieurdienst* (Engineering Service) managed by the *Verien Deutscher Ingenieure* [German Engineering Society] with funds coming from the *Arbeitsdienst,* which was in a way a type of WPA [Works Progress Administration of the

United States during the Depression]; b) from the payments of the city of Magdeburg which started on 2 February 1933 [the Magdeburg arrangements are discussed in detail below.] Nebel also had some regular *Arbeitdienst* (Nazi uniform and all) for clean-up and things like that for short periods of time."

Among the regulars were Riedel's friend, Kurt Heinisch, a baker's apprentice; Paul Ehmeyer, a jobless electrician from Austria, according to Nebel; and the jobless engineers W. Wohle and Hans Hüter. With Nebel's driving force—and in fairness, he should be given credit for this—there was quickly assembled a well-equipped team that was able to complete its first large test stand by 12 March 1931.[70]

During this time, Nebel also sought out scientific institutions as well as scientists and Governmental leaders. The purpose was twofold: Nebel wished to secure additional funds with the hope of long-term contractual arrangements and he also desired respected testimonials from Government and academia. This approach was common to all astronautical and rocket societies. Nebel recalls Interior Minister Karl Severing showing special interest in spaceflight and afterwards donating to the cause. A former Interior Minister, Alexander Dominicus, who was also the first President of the *Deutschen Luftfahrtverbandes* (German Aviation Society), attended some of the trials in the Summer of 1931 and presented Nebel with a certificate recognizing the value of his work for future high-altitude research.

Unquestionably the most prominent figure Nebel claims to have contacted was the great Nobel Prize winner Albert Einstein. An alleged introduction was obtained through Einstein's stepson-in-law, Dmitri Marianoff. The first meeting was brief and produced nothing. Then, through subsequent discussions with Marianoff, Nebel says a gathering was held on 5 May 1932 in the conference hall of the Excelsior Hotel in Berlin in which Nebel, Riedel, Einstein, Marianoff, a Professor Kapp, Professor Friedrich Simon Archenhold of the Berlin-Treptow Observatory, and "many other prominent scientists" are said to have attended. The purpose of this assembly, Nebel continues, was the formation of a scientific research-pacifistic organization known as PANTERRA, from Greco-Latin roots meaning "All Earth." Kapp was chairman, Nebel his deputy. "Our objective," Nebel writes, "was to stimulate the interest of the peoples of the Earth in the major problems of science and technology, and divert funds for armaments to peaceful and productive work. We organized this world peace program of PANTERRA." Among PANTERRA's objectives were space travel, developing atomic energy for peaceful uses, developing robots to relieve mankind of manual labor, and creating new energy sources. However, because of Nazi persecutions, Nebel says, the organization received too much adverse publicity and was soon dissolved. It is strange indeed that none of the major biographies of Einstein mention PANTERRA nor Nebel. Marianoff's own recollections of his father-in-law, *Einstein An Intimate Study of a Great Man* (1944), does devote a chapter to his numerous talks with "Captain Nebel, a rocket plane engineer," but mentions neither PANTERRA nor a meeting between Nebel and Einstein. We are thus forced to conclude that Nebel's PANTERRA story is either apocryphal or that it was a very minor event in the life of Albert Einstein.[71]

Nevertheless, Nebel genuinely used all of his energies in promoting the *Raketenflugplatz*. He was delighted one day when a man calling himself "Senior Engineer Richter" showed up in his office and offered to help in Nebel's solicitation campaign among companies and organizations. Nebel was impressed and agreed. Money came in indeed, but directly into the pockets of the so-called Engineer Richter. The VfR received "not one red pfennig," Nebel laments. Only after three months had elapsed did the Society, with great shock, realize the swindle: "Our Engineer Richter turned out to have been a former prisoner who had been convicted several times for marriage fraud."[72]

The VfR men were innate idealists and not easily disillusioned. This was the impression Einstein's son-in-law was left with after he visited. "The rocket airdrome consisted of a few starkly simple barracks and many workshops," Marianoff wrote. "The impression you took away with you was the frenzied devotion of Nebel's men to their work. Most of them were [like] officers living under military discipline. Later, I learned that he and his staff lived like hermits. Not one of these men was married, none of them smoked or drank. They belonged exclusively to a world dominated by one single wholehearted idea."[73]

The men of the VfR at Nebel's *Raketenflugplatz* managed to "prosper." That is, while almost always financially on the brink (Nebel filed bankruptcy for the *Raketenflugplatz* soon after it was established, but in a nefarious and unannounced scheme to raise money), the VfR undertook numerous experiments. Unfortunately, only fragmentary data remains.

Following the demise of the Society's journal *Die Rakete,* no regular VfR organ was issued in which test results would ordinarily be printed. Rudolf Nebel sporadically published *Raketenflug* from 1932, purportedly covering progress at the *Raketenflugplatz,* but it was not a scientific journal. *Raketenflug* was almost wholly written by Nebel and was about Nebel, or his schemes. The bulk of trustworthy information that we have on VfR experimentation comes from the chronicles of Willy Ley and the published recollections of the American G. Edward Pendray who visited the *Raketenflugplatz* on behalf of the American Interplanetary Society in April 1931. Ley's private notes record the amount of work done during the first and most active year of "official" VfR experimentation at the *Raketenflugplatz,* by May 1932: 87 rocket flights (mostly with parachutes), more than 270 static tests, 23 demonstrations for clubs and societies, and 9 "for publicity." A tabular summation of some of the flights is found in the Appendix.[74]

Ley, in his popular histories of space flight, glosses over the VfR's first flight, perhaps because this flight was unintended. On 10 May 1931 one of Klaus Riedel's watercooled *Mirak* III's simply broke loose from its test stand and went up 18.3 m (60 feet). Four days later the same rocket, rechristened *Repulsor I,* was repaired and sent aloft on its first official ascent—minus a parachute. Like school boys, the experimenters were impatient to see it fly. It too went up about 60 feet, then crashed. By 23 May one of the double-stick *Repulsors* attained a distance of more than 600 meters (1,970 ft) and 1932 improved *Repulsors* reached horizontal distances of 5 kms (3 miles) and altitudes of about 1.5 kms (0.9 miles).[75]

From a letter to Pendray, dated 2 November 1931, we get a rare inside account of VfR experimentation from Ley: "Yes, you are right, the problems of stability will be solved by the One Stick *Repulsor* (I will call it in short OSR, and our old *Repulsor* R) . . . The first idea of the R had [been thought of by Klaus] Riedel, but he didn't know how important his idea was—we only wanted to make a little flying plaything, because we had seen, we shall need long time for Mirak III. And after the first shot of the R (its price was only $8. [sic]) we saw we can work with this most simple apparatus. Meanwhile we have built 4 or 5 Rs and 3 or 4 OSRs. Now we must show it to make money . . . This series of flights will also teach us more . . . We are cooling the motors with water, ½ liter is sufficient, the fuels are O^2 and Gasoline, in the proportions of 1:2 (about). I wrote you, that we have destructed a house of the police—but now all is all right again, the president of Police—Headquarter[s] visited us and saw a shot of the OSR (the same the Ufa [motion picture company] has seen) and gave us the permission to continue our work. But the financial disturbances are hindering a little, not much, we have now no money and we had also no money before. To start the *Repulsor,* we have two rails, vertically, in a distance of 1 m from one of the houses. All operations are made through the window. This window is filled with beton [concrete] and there are only very little openings for the 'keys', as we call the instruments to open the intakes a.s.o. [and so on]. We have also had explosions with Rs, but we have also learned, to make explosions [sic] little dangerous. Look, it is only aluminum, and the chance that a piece of burning aluminum finds the way through the little openings, is very unbelievable."[76]

The incident of the crashing rocket into the police building apparently took place on 17 October 1931. Nebel, in his *Die Narren von Tegel,* also discusses it. He recalled that the one-stick *Repulsor* went up to 1,500 meters (4,922 ft) and landed 1 km (.6 miles) away. The parachute failed to function. Damage to the police building was not severe—only two tiles in the roof had to be replaced—but it was enough that the "corpulent police chief [Albert] Grezinski" called personally upon the *Raketenflugplatz* the next morning and gave Nebel the severest reprimand. A ban on the experiments was also posted. Nebel took all this in stride and cooly informed Chief Grezinski that "my greatest experiments have been conducted at this place" and that naturally there were always difficulties. Nebel says he then invited Grezinski to a demonstration. The drama was heightened by a 60-second countdown. Three weeks later, after several discussions and hearings, Grezinski's previous order to cease experimentation was withdrawn but new restrictions were posted:

1. The weight of the rocket with fuel must not exceed five kilograms (11 lbs).
2. Rocket motors (for flight) must have passed three static tests.
3. Heavier rockets required special permits.
4. Rocket flights could be made only on work days between the hours of seven and fifteen hours (7 a.m. to 3 p.m.).
5. No rocket flights were permitted on windy days.

These restrictions apparently severely affected VfR flights from there on. The deepening Depression also limited activities from 1932. Willy Ley was on the lecture circuit and found it a lot more profitable from a personal standpoint as well as for the cause of space flight. His absence also meant a loss for history of reports of the VfR's last official experiments in 1932.[77]

That winter' was severe. Apart from the weather, the economic-political situation of Germany was bleak indeed. Reparations payments were crushing enough. But with the failure of the Austrian *Credit-Anstalt* the previous year, came the financial collapse of central Europe and Germany in particular. By the beginning of 1932 unemployment was already 6,000,000 and still soaring. Hugo Hückel and other well-to-do VfR benefactors no longer felt secure enough to help the Society and soon withdrew their support. VfR membership rapidly dropped to less than 300. Desperate to continue their work on rockets, or any sort of job, the *Raketenflugplatz* regulars survived as best they could through the Government and German Engineering Society's *Ingenieurdienst* (Engineering Service) program.[78]

Young VfR member Wernher von Braun was more fortunate than most. As the son of a Weimar Republic Agricultural Minister and a founder of the *Deutsche Rentenbank* (German Savings Bank) he suffered no penury. He was as fervent as the rest in his rocketry, however, and accepted at once the offer made by the German Ordnance Department to work for their own rocket program on a secret basis with far greater facilities at their disposal. He left the *Raketenflugplatz* for the Army in October or November 1932. The others at *Raketenflugplatz* had to seek their own opportunities. Nebel presented one about this time which became known as Project Magdeburg.[79]

The Project Magdeburg Episode

Just when things were at their lowest, Rudolf Nebel discovered what he believed was a promising new source of financial support and also a means for the VfR continuing its invaluable work. With Ley on the road making his lectures and the VfR President von Dickhuth-Harrach preferring to remain in more comfortable quarters away from the *Raketenflugplatz,* Nebel was the de facto head of the testing field and proceeded on his own. He had always felt that the *Raketenflugplatz* was ''his'' anyway, having been the founder, though conveniently neglecting to consider that Riedel was a co-founder.

Nebel's solution to everything was intertwined with the new *Hohlweltlehre,* or ''Hollow Earth Doctrine,'' an absurdly pseudo-scientific theory of Peter Bender and Karl E. Neupert in which it was believed that we live inside a hollow sphere. The universe was an optical illusion. Within the center of the sphere were the Sun, Moon, and ''phantom universe.'' The latter was a dark blue sphere studded with little lights that are mistaken for fixed stars. A so-called engineer from the city of Magdeburg, Franz Mengering, also espoused this theory and maintained it could be proven if a rocket were shot up so that it would hit the antipodes, or opposite side of the earth sphere. Nebel took up the challenge. For a large fee he committed the facilities of the *Raketenflugplatz* and several of its personnel—who were only too happy to continue working on rockets for pay—toward building and flying a *manned* rocket to prove the theory. It is immaterial whether Nebel also believed the Hollow Earth doctrine. Involving the VfR in this latest scheme inevitably brought not only embarrassment upon the Society but also its downfall.

Early in October 1932, Franz Mengering spoke before prominent Magdeburg officials convincing them of the importance of testing the Hollow Earth Doctrine in their city. If proven correct, the rocket experiment would bring them the greatest ''scientific'' prestige, far surpassing the famed experiment of their early *Burgomeister,* Otto von Guericke. In 1657 von Guericke spectacularly demonstrated the force of air pressure by applying the strength of sixteen horses in pulling apart a .37 m (1.2 ft) diameter vacuated hollow hemisphere.

The rocket launch promised to reap a great tourist revenue. This argument especially appealed to the Magdeburg city fathers. They agreed to underwrite the experiment, stipulating extra police protection on ''Rocket Day'' and that ample publicity be generated. Nebel made a special trip to Magdeburg to confirm the arrangment. On 27 January 1933, a contract was signed between the leading citizens of Magdeburg, Mengering, and Nebel. For 25,000 Reichmarks the *Raketenflugplatz,* under Nebel's leadership, was to construct and launch a manned ''Pilot Rocket'' from Magdeburg on Pentecost, 11 June 1933. An additional 15,000 Reichmarks was required for organizing ''Rocket Flight Day'' as a public holiday. Pledges for the funds came from the *Deutsche Reichsbahngesellschaft* (German State Railway), the Magdeburg Magisrat, Magdeburg Streetcar Company, the Alliance and Stuttgart Insurance Company, Magdeburg Chamber of Commerce and other sources. A large-scale advertising campaign also commenced "in order to bring the greatest possible number of people to the launch place."[80]

Shortly, the President and Vice-President of the VfR, von Dickhuth-Harrach and Ley, heard of the plan and were compelled to draw up a list of charges against Nebel leading to his suspension as Secretary. Among the complaints for dismissal were that he completely neglected his secretarial duties; falsified the ledger; sold articles under false pretenses; failed to pay certain VfR financial obligations (such as compulsory medical insurance); took credit for engineering accomplishments that were never his; and brought "scientific disgrace" upon the VfR for supporting the "Hollow Earth Doctrine." Nebel was dropped from the VfR rolls but maintained his hold over *Raketenflugplatz.* As founder of the *Raketenflugplatz,* he stubbornly held that he had every right to conduct affairs there as he saw fit. The law supported him. So did several members. Magdeburg brought them work. Their support was perhaps also gratitude for his equipping and furnishing *Raketenflugplatz* and supplying living quarters. Some also held that Project Magdeburg was not a VfR activity, but a private one conducted by volunteers who happened to be members of the VfR. Herbert Schaefer, one of the participants, also stressed the need of his fellow engineers to keep on working so as not to remain idle and to keep in practice with their profession in very difficult times. Hans Hüter, another participant in Project Magdeburg, confirmed this when he wrote: "For some time, I was unable to find any work. In March 1931 I was able to get a position as assistant mechanic for the automatic long distance telephone installation. [Following a lay-off] . . . In April 1932, I joined the Free-Workers Organization '*Raketenflugplatz,* Berlin,' which was under the leadership of Diploma Engineer Nebel. Here, I worked mainly on constructive problems as well as problems of the liquid rocket and the testing and starting apparatus which were developed by the work party. Besides this, I evaluated experiments. After the flight tests requested by the city of Magdeburg with two larger rockets for 600 kg [sic] thrust [this project was] concluded with little success . . ." The charges against Nebel in the meantime could go no further and the project continued.

Other VfR members on Project Magdeburg were Riedel, Heinisch, Ehmeyer, Bermüller, Zoike, Prill and Dunst. To 21-year old Kurt Heinisch went the honor of being selected the "pilot" for the Magdeburg rocket; Heinisch was then drawing 7.5 Reichmarks in weekly welfare checks. To say he was a "pilot" is a misnomer. Heinisch's function was to merely sit in the rocket, not steer it. Almost no thought was given to his equipment or life-support system though he was provided with a parachute. The German Sunday supplements were typically vague and melodramatic about the prospects of a manned rocket flight, the newspaper *Mitteldeutschen Zeitungsblocks,* for example, sketching the following profile on Heinisch: "Restlessly active working with his rocket plans to leave then return to earth, no one dares question him about fear or failure . . . For people of his caliber there is no knowledge of fear. He worked many long years with others on the rocket and according to him had the great fortune to be selected from all the others to be the first rocket pilot for engineer Nebel's project. He is completely aware of what he is doing. At the moment of preparing for a parachute pilot's license he is currently limping from a recent jump but appears to be doing well . . ."

The manned version of the Magdeburg Pilot Rocket was to be 8 to 10 meters (26.2 to 33 feet) high, 1 meter (3.3 feet) at its widest diameter, and producing a thrust of 750 kilograms (1,650 pounds), rather than 600 kilograms given above. The fuel was alcohol and liquid oxygen. Much development work had to be accomplished within just a few months before the final rocket could be constructed. Unmanned prototypes were constructed first. Plans called for development of a 1.7/200 engine, capable of delivering 200 kgs (440 pounds) thrust for about 30 seconds with a propellant consumption rate of 1.7 kilograms (3.75 lbs) per second. The final engine was designated the 5.1/750. The Swiss-born Hüter designed most of this hardware. The latter was the largest ever made at the *Raketenflugplatz* but it was never fired because there was simply no money nor time.[81]

The Magdeburg motors were among the earliest regeneratively-cooled rocket engines known. That is, the combustion chamber was surrounded with a double wall or cooling jacket through which flowed the fuel prior to its entry into the chamber. This fuel was at first a 40% alcohol-60% water solution and later increased to 60% strength alcohol. The purpose of circulating the fuel was twofold: to cool the engine so that it was capable of long thrust durations without overheating and to pre-heat the fuel before ignition. Both fuel and oxidizer entered the combustion chamber by injectors at the lower end of the chamber. A nozzle was welded between the injectors. The fuel entered the combustion chamber by pressurized liquid nitrogen and the oxygen entered by self-evaporation.

Herbert Schaefer's daily notebook kept during this period attests to the amount of time and energy he and other *Raketenflugplatz* people expended upon the project. Schaefer himself sometimes spent 60 hours per week, both at the *Raketenflugplatz* and at Albert Schubert's private welding shop in Berlin. There were still difficult problems. The entry for 25 March 1933 reads: "Firing—Motor Exploded." From 27 March to 30 March Schaefer was at the welding shop, the aluminum welding of the motor presenting especially difficult problems as the technology was then new. In April 1933, there were 11 days of test firings Schaefer attended and six-and-a-half days at the shop. The tests were made on a new 1,000 kg (2,220 lb) thrust capacity test stand. "The sound," wrote Schaefer, "could be heard for miles." On 1 June the completed 1.7/200 prototype Magdeburg rocket was fired at the *Raketenflugplatz* satisfactorily. On the 7th the rocket left by truck for Magdeburg.

Press coverage was afforded this rocket even though it was an unmanned prototype. Nebel and his team were way behind schedule but had to satisfy the Magdeburgers that some progress was made. Normally, VfR rockets were unshrouded with the plumbing exposed in order to save weight. Because this was a public flight all the Magdeburg vehicles were covered to make them both photogenic and pleasing to the public. The metallic shroud of the first Magdeburg rocket concealed an arrangement of five tanks, two each for the fuel and oxidizer and a smaller one in the center for holding pressurized liquid nitrogen. The 1.7/200 motor weighed 3.5 kgs (7.7 lbs). The combustion chamber was of thin gauge Pantal aluminum of cylindrical shape, the top of which was a spherical dome. Total length of the motor was about 30 cm (11.8 in) while the entire length of the rocket was 280 cms (110 in) and its maximum diameter 75 cms (30 in) Dry weight was 70 kgs (154 lbs). The overall configuration was the nose-drive pattern, the same intended for the manned version.

The first attempted launch was made 9 June 1933 at 5:30 a.m. on a cow pasture at Mose, near Magdeburg. The rocket barely lifted when an oxygen valve failed and the rocket slid back, not even clearing the wooden launch rack. Few people witnessed this failure but a large crowd, including newsmen and police, showed up for the second attempt—on Pentacost Sunday, 11 June, at 11 a.m. This was "Rocket Day." The full-scale Pilot Rocket never materialized. Anxious, the Magdeburgers had to be content with a smaller, unmanned version. Another failure for Nebel and his team. A leaky gasket in the nitrogen tank prevented three-quarters of the fuel from feeding into the combustion chamber. The rocket roared for two minutes instead of 30 seconds, but never budged from the stand. On the evening of June 13, patient Magdeburgers finally saw it fly, but only to two meters (about six feet). A vent screw on the cooling cone "popped out," wrote Schaefer," and the rocket fell back, getting no fuel." These failures necessitated a fortnight's delay to completely overhaul the rocket for still other attempts.

Schaefer's notebook records trips backwards and forwards to Wolmirsstadt-Mose by bus and rail as well as six *Raketenflugplatz* test stand runs with the motor. These too were unsuccessful for the most part. A valve froze, a nozzle membrane burst prematurely, and so on. Then at 6:45 p.m. on 29 June 1933 came the final attempted shoot in the Magdeburg series. It was the highest a Magdeburg rocket ever went but still pitiful considering Nebel's promises and big buildup. Heavy rains warped the wood launch stand. It took eight seconds to clear the rack, the rocket being momentarily held back by an unaligned roller. Consequently, the rocket tilted and took off almost horizontally. Then, after 15 seconds, the rocket lost altitude, and according to Schaefer, "made a belly landing 1,000 feet [305 meters] from the rack, the motor still going full blast. It slithered for another 30 feet [9 meters]. It looked totally smashed, but the motor and the tanks were unhurt. Only the casing, fuel lines, etc., had been smashed." The ride home compounded the disaster. The truck had a flat tire in Burg, then at Glindow the vehicle lost its wheel. Schaefer and the others reached home at 4 a.m. the following morning.

Since the Magdeburg contract was partly fulfilled, Nebel received only 3,200 Marks. Project Magdeburg had grave repercussions for the Society by the end of the year. In the meantime the *Rakentenflugplatz* rocketeers continued their experiments with the left-over Magdeburg hardware. The motor and tanks were rearranged and extra tanks added so that it became a "four-stick Repulsor." This vehicle, utilizing a 1.7/200 motor, stood 2.3 meters (7.5 feet) high, was taken out to "Lover's Island" (actually, Lindwerder Island), on the outskirts of Berlin, for a launching. The "heavy rocket" ban imposed upon the VfR at *Rakentenflugplatz* in 1931 made it impossible to fly in Berlin proper.[82]

The new rocket was fitted with fins around its nose but it did not include a shroud. The launch was made on the morning of 14 July. According to Schaefer, "She rose with terrific velocity to about 3,000 feet [914 meters], suddenly tilted over up there, made a few loops, and came down in a power dive, landing some 300 feet [91 meters] from the island in the water. The parachute was ejected at the last moment before striking, thus only minor damage was sustained."

The captain who owned Lover's Island was apprehensive, not because of any potential damage but because the noise scared away his summer campers. The *Raketenflugplatz* rocketeers found an alternate spot—the deck of a motorboat on Schwielow Lake, near Potsdam. This particular launch was made on the morning of 5 August. There were no onlookers here but the river police, wearing swastikas of the recently installed Nazi regime, came by to inspect. Mostly, they were only curious but were also guarding their property, as the police owned the boat. The rocket flew to 60 meters (197 feet) but went no further because of a burst valve. It plummetted in the water. A similar flight of 1 September was likewise unimpressive, with Schaefer apparently catching a cold while searching for the remains of the rocket in the water. The Four-Stick arrangement seemed too difficult to manage so the 1.7/200 motor was now adapted to a more conventional Two-Stick pattern of two tanks in tandem. Yet the attempted boat-launched flights of 9 and 19 September of this vehicle also miscarried. Schaefer's notebook entry for the latter launch is typically brief: "19 sept mon[day] 1800 [hours] launch of the Second Two Sticker on Schwielow Lake Failure." Schaefer's notes list no other launches. This was the VfR's last flight rocket, and probably its last rocket.[83]

What of the technological "priorities" claimed by the VfR and its members in both Magdeburg and non-Magdeburg periods? The early development of regenerative cooling for rocket motors has been well analyzed by Irene Sänger-Bredt and Rolf Engel in No. 10 of the Smithsonian's *Annals of Flight*. The concept was not new but Nebel and his team gained invaluable practical experience in its use. VfR likewise gained experience with the application of liquid oxygen and alcohol as propellants. In March 1933, Hans Hüter had also constructed an interesting innovation, a roll-back corrugated protective metal shed at the *Raketenflugplatz* to protect the Magdeburg rockets from the elements. This shed anticipated the leviathan VAB (Vertical Assembly Building) constructed at the Kennedy Space Center at Cape Canaveral for housing and servicing the Saturn moon rocket.

In retrospect, the real legacy of the *Raketenflugplatz* was the practical training gained by some of its members who later served at the German Army's rocket center at Peenemünde. Herbert Schaefer puts things in perspective: "In the time frame and with the funds made available by the city of Magdeburg, the manned rocket could not have been built for a safe flight. The project, however, resulted in valuable experience with liquid propellant rockets with alcohol and LO². The application of the same funds with a more leisurely schedule to the development of a rocket with a motor of the 1.7/200 class would have been much more prudent. Even if the manned rocket had been a success, the political conditions in Germany would have made it a dead-ended effort as private industry was soon to be inhibited from any rocket development. The military had been interested in our work. While we could not join them at the time, Riedel, Hüter, Zoike and Schaefer worked at Siemens—we were 'placed in cold storage' *(auf Eis gelegt)* until we could join them when Peenemünde was organized. (I myself had gone to the USA [in 1936], and remained here, however)." Later Schaefer joined the US space agency, NASA.

After the attempted launch of the last VfR rocket from a boat on Lake Schweilow on 19 September 1933, experimentation probably ceased altogether. (Herbert Schaefer's notebook shows he visited the *Raketenflugplatz* almost on a daily basis until late 1934. Yet no static tests were recorded. There were only occasional meetings.) By late 1933 the VfR hierarchy was bitterly embroiled in the aftermath of the Magdeburg affair which had now come to a head.[84]

Nebel's improprieties, especially his using VfR resources and personnel in undertaking the absurd Project Magdeburg, led the VfR Board of Directors to take him to court. Even though the Board of Directors of the Society disavowed any connection with the Project, the Society was billed by a factory for certain parts for the Magdeburg rockets. In his confidential letter to the American Interplanetary Society of 26 December 1933, Willy Ley said the case was dismissed "for lack of proofs." Many years later, Ley explained that the District Attorney, "seeing that Nebel wore a swastika armlet, was afraid to act." Ley also wrote about a "stormy session" of the VfR at the end of 1933 in which the VfR "collapsed". The reasons were several: the terrible financial situation, strictures on private experimentation imposed by Nazi regime, and the mire of Project Magdeburg. Ley took steps to re-organize the Society by enrolling VfR members in the *Fortschrittliche Verkehrstechnik, E. V.* (EVFV—Society for Progress in Traffic Technics), originally established in 1920.

The fall of the VfR and its absorption into the EVFV was made official in an extraordinary letter by von Dickhuth-Harrach and Ley, dated 4 January 1934. The letter, sent to all VfR members, said in part:

"We are of the opinion, that the ideals and the good old tradition of the VfR shall not be allowed to perish . . . And, furthermore, we believe that the representation of these ideals cannot continue anywhere better than in the registered society, 'Progress in Traffic Technics.'

"We are therefore urging you to follow in our footsteps and help us to achieve in the EVFV that we were not able to achieve in the VfR because of Mr. Nebel's personal interest; spreading and deepening the idea of rocketry and a scientific, serious advancement of the rocket technique, without sensationalism and without unilateral commitments . . . Heil Hitler!"

Nebel's version of Project Magdeburg appeared almost 40 years later, in his *Die Narren von Tegel* (1972). Nebel does not depict it at all as a shameless episode for which he above all others deserved blame. Rather, he saw it as a great technical challenge for the VfR, "way ahead of its time," though he personally did not espouse the "false science" of Hollow Earth Theory. Willy Ley, he alleges, "interrupted our work" by unjustly stripping him of the *Raketenflugplatz* chairmanship while VfR President von Dickhuth-Harrach and others demanded private considerations for themselves. Major von Dickhuth-Harrach, for example, requested VfR engineers to modify his Opel P4 automobile in streamline form. Nebel says he turned down this request and that consequently the Major claimed one of the Society's own cars. The VfR Board of Directors, according to Nebel, stepped in, and called the criminal office which seized the Society's cash book and funds. In an "excited debate" that followed, he continues, von Dickhuth-Harrach and Ley were "excluded" from the Society and "I was again chosen Director [of the *Raketenflugplatz*] and Werner Dunst who, at the *Raketenflugplatz* kept the records, was chosen Vice-Director."

When Rudolf Nebel wrote these words, the other two protagonists had been dead for some years and could not answer the charges. Major von Dickhuth-Harrach died in 1947 and Willy Ley in 1969. Nebel himself died in 1978.

Ironically, the immediate cause for the end of Nebel's cherished *Rakenteflugplatz* was due to some leaky faucets. One day Nebel was handed a bill of 1,600 Marks by a city official for leaks that had occurred during the life of the almost rent-free *Raketenflugplatz* from September 1930 to the summer of 1934. Nebel could not pay and the lease was cancelled. The VfR files were sent to a Siemens warehouse and never turned over to the Army or the Gestapo.[85]

Post VfR Groups

Willy Ley in particular must have felt a great sense of relief at the beginning of 1934. Rudolf Nebel and Project Magdeburg were behind him. He also planned to leave Nazi Germany. Privately he wrote to his American friend G. Edward Pendray on 2 February 1934, though it was still too early to confide his travel plans yet: "Well, it was difficult and not very pleasant job the whole mess but I thought it better to do it at once. Later on it would have been even more difficult and nasty. But the 'new' Society is arranged now and I thinks its [sic] nobody left of the Nebel crowd. His name means mist or fog in English and thats what he is and what he always does . . . we'll be able to make propaganda again and to build everything anew. If possible, we'll also start co[n]structional work very soon. I already had two meetings with the engineers of the Society. Very able and experienced men."

The EVFV's propagandizing of spaceflight began immediately. In that same month appeared the organization's new journal *Das Neue Fahrzeug* (The New Vehicle), though bearing the misprinted year of 1933 instead of 1934. Willy Ley wrote the lead article, *"Die ersten Postraketen"* ("The First Mail Rockets").

This is most odd because Willy Ley claims that about this time the Goebbels Ministry of Propaganda "issued a directive to all newspaper editors that they were not even to mention the word 'rocket' in print." It is even more odd considering that *Das Neue Fahrzeug* lasted for 20 issues, from February 1934 to May 1937. An effort has been made to locate the Goebbels directive without success. However, in a letter from W. L. Schlesinger of the Astronomical Society of South Africa written to Andrew G. Haley in 1955, in answer to Haley's request for material he was gathering for a book on world-wide astronautics, Schlesinger informed him that in 1934—1935 he had been "the editor of a Jewish newspaper [and] . . . was still a member of the *Reichsverband der Deutschen Presse* [National Federation of the German Press], and, therefore received the official Nazi circulars. I remember quite vividly one of their directives to German newspaper editors; it stated that as from this date no stories about rocket research of whatever origin were to be published unless submitted, prior to publication, to official censorship. As I had written quite extensively about rockets in the year prior to the ascension of the Nazis to power, this gave me quite a jolt . . ."

This historical dilemma can only be resolved by accepting Krafft Ehricke's explanation which is that the rocket "ban" only applied to mention of military rockets as he submitted space travel articles to the journal *Weltraum* (Space) even while he worked at Peenemünde on the V-2 and also later, after being inducted in the Army to drive tanks on the Russian front. It is possible also that *Das Neue Fahrzeug* and *Weltraum* slipped by the censors or that they received clearance being judged harmless. Indeed, the EVFV's name was innocuous enough: "Registered Society for Progress in Traffic Technics." The Society's purpose also seemed harmless: "to promote traffic technology on land, water, air and space, as important means of culture through scientific investigation, popular enlightenment and fostering practical invention." So far as is known, no EVFV experimentation in any area was carried out during this period. But there is an interesting melange of articles—more or less in the popular scientific vain—in the journal. These range from Guido von Pirquet's series "On the Question of Feasibility of Spaceflight With the Means of Modern Technology" and Steinitz's "On the Stability of the Space Rocket," to von Dickhuth-Harrach's "The Destruction of the Airship 'Hindenburg,' " and Steinitz's "On The Berlin Auto Show."

Despite the Society's aim of promoting traffic technology in all forms "through scientific investigations," private rocketry manufacturing and experimentation was strictly forbidden. The severity of this edict, is shown by the four year imprisonment of Friedrich Sander, for "negligent treason" in making and selling a large order of his life-saving rockets to the Italian Government.

The most dramatic evidence of the state of affairs in early Nazi Germany and its effect upon the rocket movement is found in a letter from the young founder of the British Interplanetary Society, P. E. Cleator, to G. Edward Pendray of the American Interplanetary Society. The letter is dated 30— 31 October 1934. The closing words proved to be more prophetic than Cleator realized: "Apparently there is some trouble brewing in Germany—trouble about which Herr Ley dare not write. It seems that all my letters are opened, and their contents carefully examined. Moreover, most of the letters I get from Germany have been neatly slit open, and then gummed up. Moreover, I am requested to be very careful what I write in future, in order to, the message goes on to say, avoid trouble. I am specially requested not to use any letter heading or envelopes bearing the name of the Society. In future, all correspondence must be sent on plain paper and in plain envelopes. Finally, my letters must refer to nothing but space travel—apparently the word 'rocket' is taboo . . . I met Dr. Steinitz of the EVFV in London two weeks ago, and he mentioned nothing about it. One fact the Dr. did mention [sic] however, may have some bearing on it. I understand from him that the German Government had offered to grant the EVFV much money for research work providing the results of their research were not published in any shape or form, but were to become the property of the Government. The German Society refused the offer, knowing full well that the idea was to develop the rocket as a weapon of war . . . that is the position. It seems to be that rocket research in Germany is becoming a closed book—until the fighting begins."[86]

With the cessation of *Das Neue Fahrzeug* in 1937 and no further word of that organization, it is assumed it collapsed by that time. The end of the EVFV may not have coincidentally occurred that same year the German Army's new Peenemünde rocket center was opened. Perhaps key EVFV members joined Peenemünde's mushrooming staff.

The VfR and EVFV were not the only pre-war rocketry or astronautical organizations in Germany. There was a student group called the *Gesellschaft für Raketenforschung* (Society for Rocket Research), founded in 1927. Probably it was really not a society, but a section of the Breslau Model Club. A similar group called the *Studien-Gesellschaft für Raketen, e.V.* (Society for the Study of Rockets) began in 1928 in Frankfurt. Undoubtedly there were others.

In 1932 Johannes Winkler and his first assistant, Rudolf Engel, founded the *Raketenforschungsinstitut-Dessau* (Dessau Rocket Research Institute), but it was short-lived and was hardly a society or institute. The same could be said of the *Deutsche Raketenflugwerft* (German Rocket Flight Yard), began in 1933 in Vienna by the Austrian rocket pioneer Eugen Sänger, with two others. Later, in 1936, the *Deutsche Versuchsaustalt für Luftfarht* (German Research Institute for Aeronautics) induced him to start a *Raketentechisches Forschungsinstitut* (Rocket Research Institute) for experimentation with liquid rockets with Governmental approval. A laboratory was built at Trauen, near Luenberg, from 1937 and extensive work was conducted. But this was not a spaceflight propagandizing society so much as it was a scientific research project with potential military implications.

Another *Gesellschaft für Raketenforschung* was founded in the *Zur Rakete* guest house in Hanover on 18 November 1931 and may be considered a bonafide astronautical or rocketry society. Its founder was Albert Püllenberg, a gifted 18-year old who worked in abysmal conditions in a shed near the Hanover Airport. During one of his Army scouting trips for talented rocketeers, Captain Walter Dornberger, with Klaus Riedel, visited Püllenberg about 1935. Apparently nothing was said of the ban on private experimentation. Dornberger's account at least says that he suggested that Püllenberg take up an engineering degree. Dornberger also reported that Püllenberg was then "without assistance from anyone and was doomed from the beginning." In fact Püllenberg had several assistants in his liquid rocket development from 1933—1935. One was his friend Albert Low. Another was Konrad Dannenberg who years later worked for the United States manned space program. Even before the establishment of Püllenberg's GfR, he with others had built several ingenious *Gardienstangenraketen*, literally curtain-rod rockets. They utilized large German curtain rods as their propellant tanks. Püllenberg also established a *Raketenflugplatz-Hanover*, just like the VfR. He also raised funds by charging admission to see launches. A rocket was also displayed in City Hall. Some flights were attempted but most exploded. Dornberger still thought the young man bright. He built the Diesel-PT Rak. III, for example, which worked on cheaper Diesel fuel than gasoline. One of his rockets weighed about 15 kgs (33 lbs) empty and produced a thrust of 25 kgs (55 lbs). The sometimes loud commotion thus caused did not go unnoticed by the local Gestapo. He was called in and warned of the ban. Undeterred, Püllenberg continued to experiment both in Hanover and Bremen, and also circulated literature on space flight, until 1937. The curious GfR faded out of existence as Captain Dornberger had predicted. But Püllenberg also heeded the officer's advice. He attained his degree and joined Peenemünde in 1939. Dannenberg also joined, as did other Püllenberg "assistants."

Meanwhile, in Breslau, the cradle of the VfR, one Hans K. Kaiser, an astronomer, presented lectures on space flight from 1934—1935. On 18 August 1937 with eleven other adherents, he began the *Gesellschaft für Weltraumforshung e.V.* (GfW or Society for the Exploration of Space), an arm of the Breslau Astronomy Society. Kaiser quite rightly believed he could attract more members through the astronomy organization. He could also use the building of the larger organization for meetings. From 1938 the parent organization began publishing a journal entitled *Astronomische Rundschau* (Astronomical Review). Kaiser issued supplements dealing specifically with GfW news. The magazine itself contained both astronomical and space travel articles, including one on stratospheric balloons and high-altitude rockets by Willy Ley in the April 1938 issue. Within a very short time Kaiser's GfW dominated *Astronomische Rundschau*. In January 1939 the magazine received a new name: *Weltraum* (Space Flight). The first issue stated the journal was the official organ of the GfW. *Weltraum* too defied, or was permitted to be published despite the supposed anti-rocket ban to German editors. In fact the journal lasted until 1943, when a paper shortage prevented further publication.

Krafft Ehricke, mentioned above, and later a major figure in astronautics, was a prominent member. Guido von Pirquet and Willy Ley were honorary members. The Society also enjoyed an exchange of literature between similar groups overseas, namely the British Interplanetary Society, the Manchester Interplanetary Society, the American Rocket Society, the Cleveland Rocket Society and the tiny Peoria Rocket Society. Kaiser also lived up to his objectives and assembled an impressive space travel and astronomical library in his Breslau home which was open to all members. But in 1939 Kaiser moved to Cologne and took up a job in industry. At that time, according to a letter he wrote to the founder of the Cleveland Rocket Society, fellow German Ernst Loebell, he was working on a popular book on space flight that "should do the subject justice" with "profuse illustrations of the state of the art." He also asked Loebell, "Do you know if space travel is being treated in the coming world's fair in Rome and who is handling this matter?" Kaiser was pressing for an astronautical exhibit. Would Loebell loan some of his own models and drawings? he continued. "Aside from the world's fair," he added, "the International Transportation Exhibit of 1940 in Cologne might possibly come into question . . ." The war, of course, utterly dashed these ambitions.

Kaiser's energy still paid dividends. Cologne became the new GfW headquarters and publishing site for *Weltraum*. Ehricke was placed in charge of a Berlin section and other sections were to be found in Hanover and Munich, as well as Breslau. GfW membership rose to 400 by the opening of the war, with former VfR President von Dickhuth-Harrach becoming head of the Berlin branch. In the meantime, Kaiser, Ehricke, Fritz Schmidt and perhaps other GfW members started work at the Peenemünde rocket center.

Peenemünde seemed to loom over most all of the pre-war German rocket groups. It played a special role in the VfR story.[87]

The Military

The connection of the military with the VfR—and other pre-war German rocket societies—was to be of profound importance to the development of the modern liquid-propellant rocket. The key link in this connection was Karl Emil Becker, a Doctor-Engineer in the Artillery. His interest in rocketry pre-dated the founding of the VfR. He had studied ballistics under Professor Carl Julius Cranz at the *Technischen Hochschule* in Berlin and contributed to the 1926 edition of Cranz's famous *Lehrbuch der Ballistik* (Textbook of Ballistics) which contains a lengthy section on rockets. Becker himself may have written this part. Here was analyzed the solid fuel aerial torpedo of the late 19th century Swedish ordnance officer, Wilhelm Unge, and the 1919 paper of Robert H. Goddard. Liquid propellants were treated in a discussion of Hermann Oberth's 1923 space ships.

With the creation of the VfR in 1927 there emerged in Germany and elsewhere a great swell of publicity—much of it sensationalistic rather than scientific—which Becker and other military men could not fail to notice. By 1929 this publicity grew to such proportions, generated especially by the Valier-von Opel stunts, that Becker took direct action to start Army involvement in rockets. As a colonel and chief of the *Heeres Waffenampt* (Army Weapons Board) of the *Ballistiche und Munitionsabteilung* (Ballistic and Munitions Department), he ordered Captain Dr. Engineer D'Aubigny von Engelbrunner Hörstig (usually referred to as Captain von Hörstig) to thoroughly examine the literature to determine military potential of the liquid fuel rocket.

Much has been made of the reason for the German Army's interest in the potential of the rocket as a weapon in 1929—1930. The Versailles Treaty, which severely restricted Germany's armaments, conspicuously left out rockets. That the Versailles Treaty was honored even before Hitler's assumption of power, however, is a myth. Moments after the ink had dried upon the treaty, certain elements within the German arms industry and the military surreptitiously sought ways to contravene the armaments clause. J. H. Morgan makes a detailed study of these efforts in his *Assize of Arms—The Disarmament of Germany and Her Rearmament, 1919—1939*. There is ample evidence to show that the Versailles Treaty's silence on rockets was a factor in the early military development of the weapon. But what is not stated is the underlying motive: Germany's general move toward rearmament, and more importantly, the inevitability of the liquid fuel rocket's military development. The rhetoric of Adolf Hitler perhaps was also a factor which urged the military towards seeking a new, more powerful weapon. His constant theme was vengence over the Versailles Treaty and the forging of a new more powerful Army. "We will have arms again!" he had written in *Mein Kampf* in 1923. Whether out of vengence, feelings of military impotency created by the restricting clauses of the Versailles Treaty, or a Germanic fascination with new weaponry, the military was determined to exploit the rocket.

Von Hörstig's findings of the state-of-the-art was submitted to Becker and also forwarded to the Minister of National Defense. It is unfortunate that this document, as well as the orders that precipitated it, has vanished. Willy Ley suggests that it contained no useful technical information but only generalized and inaccurate accounts of experimenters of the day. Some of these experimenters were cranks. Neither technical colleges nor private industry were engaged in developmental work on liquid propellant engines. The VfR had not yet begun their own experiments and Hermann Oberth had failed to produce a workable liquid-fuel rocket to be fired as part of the publicity for the movie *Frau im Mond*. In short, the Army had no technical basis from which to start.

In the Spring of 1930, according to some sources, von Hörstig received an assistant in the rocket problem, Artillery Captain Walter R. Dornberger. Dornberger's own account is that he joined von Hörstig after graduating from the *Technischen Hochschule* in Berlin-Charlottenburg with a Master's degree in engineering. As he attended this school from 1 April 1926 to 1 April 1931 on a full-time basis, there is some confusion as to the precise chronology of events. In any case, after being assigned, he received a terse order from Becker: "You have to make of solid rockets a kind of weapon system which will fire an avalanche of missiles over a distance of 5 to 6 miles [8 to 9.6 km] as so to get an area effect out of it. Next, you have to develop a liquid rocket which can carry more payload than any shell we presently have in our artillery, over a distance which is farther than the maximum range of a gun. Secrecy of the development is paramount."

The first part of Becker's requirements was comparatively easy, and Dornberger eventually met the order with the development of the barrage *Nebelwerfer* rockets of World War II. The second part was an altogether imponderable task. There was no basis for the design, let alone the construction of such a rocket, and no realistic estimate could be made for its research and development. Fortunately within the Army there existed a rocket enthusiast in the right place by the name of Colonel Erich Karlewski who was partly responsible for approving expenditures for Ordnance Department experimental programs. On 17 December 1930 a crucial meeting took place in which Becker and Dornberger were present and presided by Colonel Karlewski. This was the real start of the Army's rocket program as Karlewski approved the allotment of the equivalent sum of $50,000 per year for the rocket program; an additional $50,000 was approved the following year.

The Army's *Versuchsstelle* (Experimental Station) at the artillery proving ground of Kummersdorf-West, about 27.3 km (17 miles) south of Berlin, was established for the work of making the new rockets as directed by Becker. Progress was at first exceedingly slow. In 1931, as Dornberger labored on his solid rockets, a contract was granted to Dr. Paul Heylandt's *Gesellschaft für Industriegasverwertung* (Association for the Utilization of Industrial Gases) to produce a small liquid engine capable of 20 kg (45 lbs) thrust. Heylandt's by this time was the only German industry that had experience in making such engines, the company having developed gaseous and liquid engines for Max Valier's later rocket cars. Heylandt's Army rocket motor was double-walled for regenerative cooling. Dornberger recalls, however, that the weight of the tanks and overall iron-bodied powerplant was 181.4 kgs (400 lbs), making it rather restrictive in application, apart from study purposes. Dornberger also reported that the Army Weapons Board was forced to seek out individual rocket inventors and to secretly "support them financially, and wait results." No progress was made in this effort so that "we had therefore to take other steps."[88]

In the Spring of 1932 three men in mufti arrived at the VfR's *Raketenflugplatz*. These visitors were Colonel Becker and his two staff officers, Captains von Hörstig and Dornberger. In this first direct contact with the VfR, Colonel Becker and his men only hoped to obtain whatever technical information they could. They had dressed in civilian clothes in order to arouse as little attention to themselves as possible. By this time the *Raketenflugplatz* was a well-publicized concern that often received visitors. But Dornberger and his companions cared not the slightest for space flight. "The value of the sixth decimal place in the calculation of a trajectory to Venus interested us as little as the problem of heating and air regeneration in the pressurized cabin of a Mars ship" he recalled. "We wanted thrust-time curves of the performance of rocket motors. We wanted to know what fuel consumption per second we had to allow for, what fuel mixture would be best, how to deal with the temperatures occurring in the process, what types of injection, combustion-chamber shape, and exhaust nozzle would yield the best performance."

However, it was not possible to get such data. Dornberger inferred that despite the many experiments conducted, the *Raketenflugplatz* lacked proper scientific means of measuring rocket performance, though dynamometer read-outs were available by the Spring of 1933. These gave thrust-time curves and fuel consumption rates but not temperatures. Pressure data were always poor.

The VfR of course had no money to pay for the equipment that was needed. Surreptitiously, Rudolf Nebel had approached the Army to solve the financial problems. The Army was unimpressed with the VfR's rockets which they considered playthings, but did consent to a demonstration at Kummersdorf. Nebel kept this arrangement secret from the VfR Board of Directors though the young and most talented of the *Raketenflugplatz* rocketeers, Klaus Riedel and Wernher von Braun, were necessarily told as Nebel took them along to perform the experiment. Nebel, supposedly representing the VfR, was to get 1,360 Marks, contingent upon a successful firing.

The Army required that the VfR rocket eject a red flare at the peak of its trajectory, in order to track it with phototheodolites and ballistic cameras. Nebel selected a large one-stick Repulsor, 3.6 meters (12 feet) long 101.6 cm (40 in.) in diameter, weighing about 20 kgs (45 lbs) loaded, and producing about 60 kgs (130 lbs) thrust. The alcohol fuel was fed by compressed nitrogen and the motor cooled by a water-filled jacket. According to Dornberger: "I remember the great disappointment in August 1932, during a demonstration at Kummersdorf, when a rocket of this type built by the *Raketenflugplatz* group, after rising vertically for 100-odd feet (30.4 meters) sharply swerved into a horizontal course and crashed in a nearby forest."

Nebel did not get the large Army contract he had hoped for. Becker and his staff were willing to support serious liquid rocket development work, but only in secrecy and on their facilities. Nebel could not accept these terms.

Army support did come, but not as Nebel or the others had expected. During his visits to the *Raketenflugplatz* and at the test at Kummersdorf, Dornberger was "struck" by the energy, shrewdness, and "astonishing theoretical knowledge" of von Braun. "It seemed to me," he also noted, "that he grasped the problems and that his chief concern was to lay bare the difficulties." Becker provided von Braun with a research grant to simultaneously pursue his rocketry work and continue his education in physics at the University of Berlin *(Friedrich-Wilhelms-Universität)*; his rocket research was now conducted in secrecy at Kummersdorf.[89]

On 1 November 1932 (some say, 1 October) von Braun became the first VfR member employed in the German Army's rocket program. Heinrich Grünow, described as "a genius mechanic," was the next to join. From 1937 when the program shifted from Kummersdorf to Peenemünde, several more VfR men were added to the roster. All had been working at Siemens Halske since the dissolution of the Society. They included Klaus Riedel, Kurt Heinisch, Helmut Zoike, and Hans Hüter. At Siemens they had not entirely divorced themselves from rocketry. According to Zoike: "This activity [at Siemens] was viewed by us as a continuation of our rocket activity, since at this time we considered ourselves to be able to design and build missiles but were fully aware of our limitations in the guidance and control field. Captain Altvater, a friend and mentor of Klaus Riedel, was head of the activity there which was the development of a three-axis autopilot for aircrafts [sic]. This included all kinds of gyros and instrumentation as well as control systems that we might later use for future missile applications. I worked first in the laboratory, and later in the flight-testing groups at Tempelhof, and later at Marienfelde and Schoenfeld. After a short eight weeks activity in the German Army, I then joined Dr. Wernher von Braun's group on 1 September 1938 at Peenemünde."

Hermann Oberth also joined the program at Peenemünde though it was a complex and drawn out process. Oberth was Rumanian and had to wait for the slow-grinding bureaucracy of the German Foreign Office before he was granted German citizenship. By that time, recalls von Braun, "the V-2 rocket was practically completed and mass production was just about getting started." Oberth was placed on the staff nonetheless and wrote "an excellent paper on the optimum weight in multistage rockets." Oberth also reviewed Peenemünde's supersonic wind tunnel techniques and afterwards submitted a proposal for a solid-fuel antiaircraft missile which was started but never completed.

Oberth was the unquestioned elder statesman of space travel theory and by far pre-dated the others in his rocket studies. But von Braun was the superior engineer and administrator as well as the first VfR man hired. He was Dornberger's first choice as a technical assistant. He thus became the top ranking civilian of the rocket program and virtually the chief rocket scientist, responsible directly to Dornberger who was elevated to a Major General.

The question of "morality" has often been asked of von Braun and his VfR colleagues, in committing themselves to the design and construction of what became one of the war's most awesome weapons. Von Braun very simply wished to build bigger rockets with space travel always in mind. "I was sure Reinickendorf [the site of *Raketenflugplatz*] was utterly inadequate even to commence the vast experimental program which must be the precursor of success," he wrote. "It seemed that the funds and facilities of the Army were the only practical approach to space travel." Von Braun's explanation is entirely plausible, especially considering the financial straits of the Society during its final days. From the personal standpoint when he was hired by the Army late in 1932, he thought of nothing but rockets and wished to persue his education as far as possible, preferably along the lines of his avocation. The Army obliged. He consequently attained his doctorate in physics in 1934 at Army expense, his thesis covering theoretical and experimental aspects of liquid propellant rocket engines. There was also during the pre-Hitler time no inkling where the work would lead. "It is, perhaps, apropos," he said, "that at that time none of us thought of the havoc which rockets would eventually wreck as weapons of war." Gerd de Beeck, an illustrator at Peenemünde, was so infected by the space travel talk of the old VfR and other space travel society members, that he was induced to render his own version of the *Woman on the Moon* on the side of an experimental V-2.

From the military-political perspective, von Braun and his team were fulfilling a duty to their country. In time of war he served as fervently as Goddard during World War I when the American was experimenting with a rocket for military applications and tested it before U. S. Army officers at the Aberdeen Proving Grounds in Maryland just before the Armistice. In the Second World War several members of the American Rocket Society similarly applied their knowledge in the defense of their country.

Standing 14.3 m (46 ft 11 in) high, 1.6 m (5 ft 5 in) in diameter, and weighing 12,805 kgs (28,229 lbs) loaded with a sea-level thrust of 27 tons (59,500 lbs) and capable of a 320 km (200 mile) range, the V-2, or Vengeance Weapon 2, was indeed a terror. But the small Baltic coast village of Peenemünde which had become the site of the first long-range rocket was also the site of the beginning of the spaceship. Von Braun and the other space enthusiasts easily recognized this and conveyed their jubilation to the others. General Dornberger who had so blatently dismissed the prospect of a flight to Venus or Mars when he first visited the *Raketenflugplatz* became a convert. Following the first successful launching of a V-2, he told his staff, "This third day of October 1942, is the first of a new era in transportation, that of space travel . . . So long as the war lasts, our most urgent task can only be the rapid perfecting of the rocket as a weapon. The development of possibilities we cannot yet envisage will be a peacetime task. Then the first thing will be to find a safe means of landing after the journey through space . . ."[90]

Almost 15 years exactly, on 4 October 1957, the first artificial satellite *Sputnik I* was lofted in orbit. Its means of launching was based in large measure upon the knowledge that the large-scale liquid-propellant rocket was workable as proven by the V-2. Wernher von Braun and his team were not surprised. They had begun their careers long before in an outlandish organization called the Society for Space Ship Travel.

V

The Russians

GDL and GIRD

Prior to the launch of *Sputnik I,* almost nothing was known of Russian rocketry and astronautical organizations of the 1920s and 30s. Not only were these activities still veiled in secrecy, particularly the military projects, but there had been no need to propagandize the early efforts. The emergence of astronautical history committees and symposia of the post *Sputnik* era radically altered this situation. Memoir papers were presented and never-before-seen photographs revealed to the West. Though much is still unknown, particularly military-political complexions of the rocket groups and personal interactions, it is now possible to appreciate the enormous efforts that were made contemporaneously with their counterparts in the West. It is unfortunate that the bulk of the Russian researches was hidden. How differently developments could have turned out had there been a cooperative pooling of technologists, talents, energies, and funds towards the accomplishment of mutual goals.

After the closing of the First World Exhibition of Interplanetary Machines in June 1927, there existed several loosely-knit groups of astronautical enthusiasts in Moscow, Leningrad, and other large Soviet cities. It was only a matter of time before they set up formal organizations. Leningrad, not Moscow, was to be the first focus of these unions. Professor Nikolai Alexeyvich Rynin, an indefatigable disseminator of the space travel idea, was responsible for one of the groups. In late 1928 he had succeeded in bringing together a "Section of Interplanetary of Travel" of the Leningrad Institute of Railway Engineers (now the Obrastove Railway Engineering Institute). Rynin, who was the dean of the Air Communications faculty of this institute, was elected Chairman. Its membership included fellow instructors, engineers and students. This may have been the identical group, or one that led to, the "Rocket Research Section" of the Leningrad Institute of Communication Engineers, which in 1929 was headed by Rynin with the assistance of K. E. Veiglin and Yakov I. Perelman. Perelman's own prolific contributions up to this time have already been discussed. K. E. Veiglin is more obscure, though Rynin credits him with being one of the earliest exponents of interplanetary flight. Veiglin wrote an article "*Sverkhaviatsiya*" (Super Aviation) in the journal *Pirod i Lyudi* in 1914.

This group did not stop at publications. From 1929 they began modest experiments with small pyrotechnic rockets with the intention of gradually increasing the dimensions and charges to reach the "stratospheric" rocket stage, i.e., rockets capable of ascending to 100 kilometers (62 mi) or more. Use of liquid fuels were considered. Soviet cosmonautic historians say little else of this section, or sections, and it is inferred that their labors led to naught; that is, with the notable exception of assisting Rynin in the preparation of Volume 5 of his astronautical encyclopedia, the volume entitled "Theory of Rocket Propulsion." Rynin also suggested a national or international research institute of interplanetary travel in 1929 but this was too premature or impolitic. (Not until 1950 was the First International Astronautical Congress convened and even then, the Russians did not join.) The other organization, which actually preceded Rynin's Institute of Railway Engineers group, was the Gas Dynamic Laboratory and which proved of greater significance.[91]

The origins of the Gas Dynamics Laboratory had a long "pre-history," going back to 1894 to the military powder rocket experiments of the chemical engineer Nikolai Ivanovich Tikhomirov. Eighteen years later Tikhomirov submitted his ideas to a military panel headed by the great aerodynamicist Nikolai Yegorovich Zhukovsky. In 1916 a positive report was received but it was not until after the Revolution and subsequent Civil War that funds were finally granted to Tikhomirov, then 60 years old, to set up a laboratory in Moscow with an adjoining 17-lathe shop for making and testing the Soviet Union's first "smokeless powder" war rockets. This facility was called the Laboratory for Development of Engineer Tikhomirov's Invention. From 1922 the organization began to operate partly in Leningrad and to branch into airplane take-off rockets. A complete shift to Leningrad was made in 1925, and in June 1928 enlarged facilities were added, along with a new name: The Gas Dynamics Laboratory, or simply, GDL.[92]

Boris Sergeyevich Petropavlovsky, an artillery officer, was made director of the new organization and continued until 1932. A test stand was established in the *Petropavlovsky Kepost* (Peter and Paul Fortress), and drafting offices were moved into the second floor of the Admirality building south across the Greater Neva River. In April 1929, one of the brighter staff members, 21-year-old Valentin Petrovich Glushko, who had recently graduated from Leningrad University, suggested an expansion of GDL's program by instituting a liquid and electrical rocket engine subdivision. The plan was accepted and young Glushko was placed in charge of it. The new sub-section was officially inaugurated on 15 May 1929 and became Department II of GDL. Department I kept up the original solid propulsion work.[93]

With the founding of Department II the Soviets began liquid-fuel rocket development. Soviet histories of the GDL are rich in details of the great technological accomplishments of the group, but say almost nothing of the applications to space flight of this impressive hardware. Officially the overall organization was called the Gas Dynamics Laboratory of the Military-Scientific Research Commission of the Revolutionary Military Council of the USSR. In this respect it was a scientific research organ of the Red Army and in no way was comparable to the civilian VfR and other true astronautical societies in the West. Indeed, as the head of the Ordnance Department of the Red Army, Deputy People's Commissar for Army and Navy Affairs, and Vice-Chairman of the USSR Revolutionary War Council, General (later Marshal) Mikhail Nikolayevich Tukhachevsky was the supervisor of the GDL and was directly responsible for allocating its funds. In a letter to the commander of the Military Engineering Academy of the Red Army in 1932, Tukachevsky had written: "The liquid jet reaction engines recently designed at the GDL will be of great importance to the future." This future was the military. Section II designed rockets for the Army or Air Forces, either as weaponry or as take-off devices for aircraft or rocket planes. The electrical rocket engine stands alone as a unique exception. This type of engine produces an extremely small but sustained thrust and could only be used for space flight. At that time the engine had no military applications. Two possible answers for this research are that it was both a purely exploratory project and a pet diversion of the leader, Glushko.

Glushko, like von Braun in Germany, had been smitten early in his life with the space travel dream. This dream never died, even when researching rocketry for military purposes. Glushko was one of the first to describe artificial satellites and space stations in his article "Stansiia vne Zemli" (Station Beyond the Earth) appearing in the Leningrad popular scientific journal Nauka i Tekhnika (Science and Technology) for 8 October 1926. The author was then 18.

As for the electrical rocket engine per se, Glushko conceived an electrically propelled space-ship named the "Helio Rocket Plane" in 1928—1929. It consisted of a hollow sphere with a series of electrical rocket engines mounted circumferentially. The electrical energy was fed in by a giant mirror or "thermoelement" surrounding the sphere and facing the sun. Goddard jotted down notes on reaction by ion streams as early as 1906, and with two students experimented with "electrified jets of gas" in 1916—1917. Glushko was probably unaware of these developments, though it is interesting to note that he was inspired by the work of American astrophysicist John A. Anderson in 1922—1926 with the "exploding [i.e., vaporizing] of metal wires to study high temperature spectra." Anderson was a member of the Mount Wilson Observatory in Pasadena, California, and had helped Goddard in the Summer of 1918 with his researches on solid-fuel (smokeless powder) multiple charge rockets.[94]

GDL's electrical propulsion work led nowhere though it does represent the world's first electric rocket. GDL's liquid work produced more positive results but is not dealt with here because of its usually wholly military and non-space character. Nonetheless, GDL pioneered in a new form of space propulsion and also set the scene for later Soviet rocket organizations. Many years after GDL also became a bonafide space hardware producer. Glushko was still one of its leaders.

Two new Soviet rocket organizations were started in 1931. These were the Gruppa po Izucheniyu Reaktivnogo Dvizhenia (GIRD), or Group for the Study of Reaction Motion, in Moscow and the Leninradskaya Gruppa po Izucheniyu Reaktivnogo Dvizhenia in Leningrad (LenGIRD), or Leningrad Group for Study of Reaction Motion. Popularly, the former became known as Mos-GIRD to differentiate it from its brother organization in Leningrad. The founding date is ascribed as 18 November 1931 for MosGIRD. The beginning of LenGIRD is not entirely possible to date precisely because, as we shall see, some preliminary work had already been undertaken prior to its "official" founding. Both came under the all-encompassing umbrella Society for Assisting Defense and Aviation and Chemical Construction in the USSR (Osoaviakhim). Ostensibly, the Osoaviakhim was, in the words of the Soviet writers Raushenbakh and Biryukov, "a voluntary society responsible, among other things, for aviation and technical sports and for supporting the construction of gliders and sports aeroplanes . . ." Actually it was a para-military organization that was "voluntary" in the Soviet sense and did as much as it could to create a national appetite for aviation, particularly among Russian youth. This included managing gliding schools, civil defense exercises, and technical training for wireless operators and machine gunners. Nothing is known of the "chemical warfare" part of *Osoaviakhim's* name and no Soviet gas-warhead rockets seem to have been made. If anything, the English rendering of "scientific warfare" is a more appropriate translation. The later rockets fit in this category; though at first, rockets were considered as sport or flying hobbies, with potential military overtones.

The members of the GIRD's were young, such as Sergei Pavlovich Korolev, the most famous of the early Soviet rocket designers and who himself became head of the group. On the outset, LenGIRD and MosGIRD were less militarily inclined than GDL, but were later funded by the military. Tikhonravov, one of the great early pioneers, subtly observed that, "it should be noted that the greater part of GIRD's funds was obtained not only through the efforts of its leaders but also because these efforts met with complete understanding on the part of M. N. Tukhachevsky." Even so, sufficient funds were lacking. Tsander's biographer speaks of "miserly salaries" received by Tsander and his cohorts at MosGIRD. MosGIRD was also humorously referred to as the *Gruppa inzhenrov, rabotayshchaya darom* (Groups of engineers working gratis).

Still, with steady governmental subsidies, Tsander and other dedicated pioneers could be assured of working full time on their all-consuming rocketry, once the groups did get under way. There were also teams of technicians and specialists to whom they could turn. GIRD became a sort of superstructure. There were other GIRDs elsewhere, with MosGIRD as the national head-quarters. MosGIRD therefore shortly became known as Central GIRD, or CGIRD, though the terms CGIRD, MosGIRD, and GIRD are used interchangeably. MosGIRD's headquarters were modest at first, in a dark, damp but spacious basement (a former wine-cellar) at No. 19, Sadovo, Spasskaya Street. After cleaning, whitewashing, wiring, and the installation of two obsolete machine tools, the premises began to hum with activity. From here MosGIRD and later CGIRD, directed not only their own extensive program but also established contact with the other GIRDs as far away as Kharkov, Baku, Tiflis, Arkhangelisk (Archangel), Novocherkassk, Orenburg, Dniepropetrovsk, Kandalaksha, and Bryansk. Nothing, however, has come to light of these other GIRDs, except that in Kandalaksha, near the Arctic Circle, a homemade rocket was launched in 1935.[95]

Widespread knowledge within the USSR of MosGIRD and the diversity of its activities seem evidenced from the available contemporary literature. The Moscow journal *Tekhnika* (Technology), for example, published a collection of congratulatory letters a few months after the formation of the organization. Emphasis was not on the military, but upon the exploration of the stratosphere. A. Bikchentov, a "working correspondent" of Archangelisk wrote: "I will be an active struggler for the realization of the problem of applying reaction engines for conquering the stratosphere." A. Boiko, an electrician of Odessa, said, "The communication on the creation of GIRD made me very happy. At the time when investigations abroad are moving ahead upon the rocket problem, we only now begin to devote due attention by our own scientists. On my part I will be glad to accept participation in this work." N. Akulov of Kharkov noted that the "Realization of rocket flying opens a large space for scientific exploration of the upper layers of the atmosphere as well as in the region of practical applications." And from still other cities came offers. Engineer M. Abelev of Leninakan wrote, "Reactive engines and rocket method of flying interest me for three years already. Please furnish me the literature and give me the necessary directives for work." The Ukranian Hydrometeorological Institute of Kiev also informed the editors and readers of *Tekhnika* that a section of "aerology" had been started.

On the popularization side were also letters of support from Tsiolkovsky himself, from Yakov Perelman, from Willy Ley in Berlin, and from the writer Anatoly Glebov in Moscow, who wrote: "My state play 'Gold and Brain' was a success in 1927—1928 in the Zamoskvoretsky Theatre, the Latvian theatre 'Skatufe' and in 1930 in Erfurt, Germany, under the name '*Raketenflugzeug I.*' In my latest play 'Morning,' shown at the Revolution theatre, I likewise touch upon rocket flying. I am always ready to be useful to you in the line of artistic propaganda."[96]

The events surrounding the start of both MosGIRD and LenGIRD were not so formal as would first appear. Depending upon the validity of the biography of Korolev by Yaroslav Golovanov and the memoir account of the first chairman of LenGIRD, V. V. Razumov, the two groups began independently of one another, but later became part of a national network of GIRD. MosGIRD appears to have had the edge in the chronology. Fridrikh Arturovich Tsander was the moving force. By the later 1920s he had completed his first *Opytnaya raketa* or Experimental Rocket No. 1, (OR-1). It was modified from a blowtorch obtained from the factory where he worked. It used air under pressure with gasoline and was ignited by a spark plug. A gauge-cock adjusted the fuel consumption and a disposable conic nozzle ensured exhaust velocities exceeding the speed of sound. By stringing up the OR-1 on two wires over a balance scale he could measure thrusts and temperature ranges for different fuel-oxidizer ratios. Tsander had been very closely associated with the Society for Interplanetary Communication (OIMS) during its brief life in 1924. He had worked for the establishment of a similar "Interplanetary Sub-Section" within the Sports-Aviation Section of *Mosaviakhim,* the Moscow branch of the *Osoaviakhim.* This effort failed. By 1930, with a motor in hand and letters of approval from the distinguished aerodynamicist Professor Vladimir Petrovich Vetchinkin, Tsander started his campaign afresh.[97]

Thrust measurements on the OR-1 were miniscule. They amounted to 145 grams (5 ozs) on the average. Tsander nevertheless drafted a follow-up program for the testing of both liquid oxidizers and liquid propellants as well as new alloys. Like so many other rocketeers, his ambitions far exceeded his budget. His dream was a fully-equipped "jet rocket testing station." This would be the beginning of the space-ship. Always present in his mind was also a reborn OIMS. Thus, he again turned to *Osoaviakhim.*

In addition to inculcating an interest in flying among Soviet youth (much like Hitler Germany supporting gliding before the war to breed a new generation of pilots for the *Luftwaffe*), the *Osoaviakhim* also provided financial assistance to promising inventors. Tsander had already approached other likely supporters but without luck. Occasionally there had been encouragement. From one N. K. Fedorenkov—unfortunately, we know nothing about him—on 12 May 1931, came the following: "Dear Fridrikh Arturovich! On May 10, I visited the administration of MOLA [Moscow Society of Astronomy Enthusiasts], where the question was raised of the creation of a section in MOLA . . . they considered the creation of a section expedient and will send their opinion to the Central Council of Osoaviakhim . . . I suggest that you contact MAI [Moscow Aviation Institute] and VAI [All Union Aviation Institution] and invite them to join with you in creating a society similar to that which existed in 1924. On the other hand, we should first get in contact with *Osoaviakhim.* The position which has arisen requires great effort and energy for the creation of the Society for Study of Interplanetary Voyages with centers in Moscow and Leningrad, with departments throughout the Soviet Union."[98]

Enter Sergei Pavolovich Korolev. This 24-year-old gliding devotee, who became the leading space ship designer of the Soviet Union a generation later, met Tsander in 1931. Whereas Tsander's passion was the development of the rocket for man's exploration of the universe, Korolev during these years was wholly wrapped up in ever faster and more efficient aircraft. Since 1927 he had been a worker in the aircraft industry and in the following year had already advanced as the head of a designer brigade in one of the industrial All-Union Aeronautical Institutions. All the time he was furthering his education, and in 1929 graduated from the N.E. Bauman High Technical School in Moscow. His thesis was on the SK-4 airplane (the initials standing for his name and the numeral indicating it was the fourth aircraft he had built). His instructor-supervisor in this project was a giant in Russian aviation, Andrian Nikolayevich Tupolev. In the same year in partnership with another titan in Soviet aeronautics, Sergei Vladimirovich Ilyushin, Korolev constructed the glider *Koktebel* (named after the town) which stayed aloft for a record time of four hours and 19 minutes at the Sixth All-Union Glider Competition. But by 1930 a new interest had leaped into his mind: space travel. Golovanov catalogs the possibilities of how Korolev may have first thought about space flight: the writings of Tsiolkovsky; the space exhibits at Kiev and Moscow in 1924 and 1927, respectively; the Tolstoy movie, *Aelita* in 1924; the close proximity of Mars to Earth that year; the sensationalistic stunts of Valier and von Opel in Germany; *Frau im Mond;* the Goddard "Moon rocket" publicity; and the celebrated slide talks of Tsander. However it happened, Korolev became a life-long convert by 1930.[99]

Korolev never kept a diary and his initial meeting with Tsander also cannot be dated. Certainly when he became a senior engineer at the TsAGI, the *Tsentral'nyi Aerogidrodinamicheskii Institut* (Central Aero-Hydrodynamics Institute) in Moscow, he became good friends with Tsander. Korolev joined TsAGI in June 1930, and Tsander began working there in March 1931. Their approaches to the great problem of space flight, they soon discovered, differed widely. Tsander was the visionary. Korolev believed the most attenable solution was the mating of rockets to airplanes. This was the path chosen by Valier and von Opel.[100]

For a functioning motor, Korolev did not have far to look. Why not a modified version of Tsander's OR-1? It was the only working "reaction engine" available to his knowledge. Because GDL activities were secret he could not have known that by the time Glushko's Department II had constructed the Soviet Union's first liquid rocket engines, the ORM *(Optynyy Raketnyy Motor)*, and the ORM-1. The ORM-1 is actually considered the USSR's first workable engine and produced 20 kilograms (44 lbs) of thrust when burning liquid oxygen and gasoline; it could also work with nitrogen tetroxide and toluene. Korolev would later benefit from this groundwork, but for now he and Tsander would have to lay their own. Obviously Tsander's OR-1 was inadequate and after discussions a whole new system was decided upon, the OR-2. The undertaking, Korolev felt, should not be amateurish and required both approval and the help of the State. Korolev would take the plan to *Osoaviakhim*. In this the protege was more effective than the master. Perhaps the positive outcome was due to the requisite qualities that Tsander lacked—Korolev's capacity for organization and most of all, his pragmatism.

Through Korolev's intercession on 18 November 1931, an important document in the history of Soviet rocketry was signed. It was a contract between the *Osoaviakhim* and Fridrikh Arturovich Tsander. It reads, in part, as follows: "Socialist Agreement on Strengthening the Defense of the USSR, No. 228/*10, 18 November 1931. We, the undersigned, being on the one part the Chairman of the Bureau of Aviation Engineering of the Research Department of the Central Council of Osoaviakhim* of the USSR . . . witnesseth that Comrade Tsander agree to perform the following: (1) Planning and development of working drawings and production of an experimental reaction engine, the OR-2, for a reaction airplane, the RP-1, namely: a combustion chamber with a de Laval nozzle, tanks for fuel with safety values, a tank for gasoline, to be completed by 25 November 1931 . . . (2) Calculation of combustion chamber by 2 December 1931. Testing of tanks for liquid oxygen and gasoline by 1 January 1932. Installation on an airplane and flight testing by the end of January 1932 . . . For work performed, Comrade Tsander will receive an award of 1000 rubles, to be paid (if the work is performed) at the beginning of the work on 20 November 1931, and completion of the work, 500 rubles each date . . ."[101]

MosGIRD was actually formed before this contract. Golovanov sums up this paradox: "All of the former colleagues of the Moscow GIRD unanimously affirm that it is difficult to determine the precise date of its formation, however paradoxical this may seem, since GIRD began working not only long before the date of the order that officially formed it, but even long before the basement was found . . . But since history loves precise dates, we must state that the first documentary mention of the organization was dated 20 September 1931 . . ."

This was a letter from the GIRD secretary to Tsiolkovsky informing him of the organization's establishment: "In Moscow, in the Bureau of Air and Technology of the Scientific Research Sector of *Osoaviakhim* . . . a group for the study of reaction engines and reaction flight has finally been created. I am the responsible secretary of the group, which, incidentally is called GIRD." The *official* order establishing GIRD is dated 14 July 1932. Korolev was designated its chief as of 1 May 1932, but had already been long in command.[102]

The rocket plane was first on MosGIRD's agenda. Tsander began working on the OR-2 in September—October 1931, before the *Osoaviakhim* agreement had been drawn up. It was to produce 50 kilograms (110 lbs) of thrust on a mixture of liquid oxygen and gasoline pressure-fed into the combustion chamber by nitrogen. Circulating water through a closed circuit cooled the motor. While Tsander was busily engaged on this part of the project, some tests of which were conducted in a derelict German church outside Moscow, Korolev collaborated with the senior glider designer B. I. Cheranovsky to adapt the motor to Cheranovsky's tailless BICh-11 plane. Its configuration was of the "flying wing" type with the wings fabricated of plywood and spanning 12.1 meters (39.7 ft). The plane's length was 3 meters (10 ft), height 1.2 meters (4 ft), and weighed 200 kilograms (440 lbs), exclusive of engine. Yet the RP-1 *(Raketoplan-1)* never flew under power. Nine unpowered flights were made, some with the motor installed.

Tsander was not destined to see his beloved engine operate. It was completed on 23 December 1932, its builders becoming the recipients of a special certificate of honor. The schedule set by *Osoaviakhim* could, however, not be met. Combustion tests began 18 March 1933 in the vicinity of Moscow, but the inventor-dreamer Tsander was absent. He had gone to the mineral spa of Kislovodsk because of overwork. Before departing, he had designed a more powerful engine of 600 kg (1,322 lbs) thrust, as well as three versions of a five ton thrust engine. This project too he did not realize. Worn out and racked with typhoid fever, which he contracted on the train to Kislovodsk, Tsander died 28 March 1933 in his 46th year.[103]

Flight trials with the RP-1 with its motor installed were made, but without the engine fired up. After GIRD and GDL were merged, says Shchetinkov, one of the early GIRD technicians, "the work on this rocket aircraft stopped because of the glider was worn out.'' In the meantime, MosGIRD (or GIRD) not only branched out into many other spheres, but also strengthened in size and structure. Korolev early recognized that the fitting of a rocket engine to a glider was not so elementary as first appeared. Most of all it required skilled technicians. He therefore set about both recruiting and training personnel with classes taught at *Osoaviakhim* by Professor B. S. Stechkin and others. GIRD thus came to offer the world's first courses in reaction propulsion. This pedagogical development was not to happen in the United States until the late 30's at the California Institute of Technology.

Korolev also advocated a GIRD periodical, *Sovetskaya-Raketa*. Like its planned precursor *Raketa* of 1924, however, it never materialized. Golovanov indicates that by 1933 there was a GIRD ''wall-newspaper'' called *Raketa*. Probably by the term ''wall-newspaper'' is meant a strictly internal organ or newsletter posted to a bulletin board and not for public dissemination. News of the Society and its aims were circulated in other ways. One came from an unexpected source. Out of the remoteness of provincial Kaluga, Konstantin Tsiolkovsky advertised the creation of the group by reprinting a letter from GIRDer Ivan Petrovich Fortikov in his *The Semi-Rocket Stratop-lane* (1932).

By this year also, the enlarged GIRD had already taken shape. Four teams were organized. Overseeing them was a technical council, or Board of Directors. Korolev was chairman. Other members were Tsander, Mikhail Klavdiyevich Tikhonravov, Yevgeny S. Shchetinkov, Leonid Konstantinovovich Korneyev, Yuri Aleksandrovich Pobedonostev, A. V. Chesalov, Nikolai Ivanovich Yefremov, and N. A. Zheleznikov. The teams were staffed by one of the Council members. One gathers that Tsander's position on the governing Council was largely honorary, young Korolev dominating the organization. The first team was headed by Tsander, the second by Tikhonravov, the third by Pobedonostev, the fourth by Korolev. The first team was assigned the development of liquid engines and flight rockets overall.

Team two was assigned: Project 03, the RDA-1 engine with pump-fed propellants, designed for rocket aircraft; Project 05, a flight rocket utilizing the nitric-acid kerosene ORM-50 engine designed by the GDL; Project 07, a liquid oxygen/kerosene engine flight rocket; and Project 09, a hybrid flight rocket. Pobedonostev's third team concerned itself mainly with ramjets, pulsejets, and solid-propellant projectiles. Korolev, was charged with his first love, rocket airplanes. This meant not only the RP-1, but its successor, the RP-218 two-seater experimental rocket aircraft with a cluster of three nitric acid/kerosene engines producing a total thrust of 900 kilograms (1,980 lbs). There were also many other projects and a swelling staff of design chiefs. By the end of 1933, as estimated by Golovanov, about 100 individuals worked for GIRD. Some of GDL's projects were also taken over by the Moscow group and this also meant military involvement. It also meant scores of rockets and rocket engines. The history of Soviet rocketry and rocket groups in the 1930s is therefore an inordinately complex one that can only be barely sketched here.[104]

Throughout its first year of life, MosGIRD (or GIRD) was civilian, voluntary, and free of any Army invention even though it was under the aegis of the para-military *Osoaviakhim*. In 1932 the situation changed. The organization soon came under the scrutiny of General Tukhachevsky. He had been appointed Deputy Chairman of the Revolutionary Military Council on 19 June 1931 and was in charge of armaments for the Red Army. The GDL thereby came under his immediate jurisdiction. Soon after hearing of the nascent GIRD, according to Golvanov, Tukhachevsky ''realized that this new affair should be supported, but it was more difficult to take it under his wing than it was with the GDL. The Gas Dynamics Laboratory had developed historically as a military organization . . . GIRD was rooted in *Osoaviakhim* and it wasn't quite proper to rob *Osoaviakhim* of a group that it had only just set up. Tukhachevsky had a different plan: not to take GIRD away from *Osoaviakhim,* but to change its sign and to set up a real scientific center on its basis.''

The General accomplished this through direct negotiations with Korolev. Tsander also agreed with the idea of unification of GIRD with GDL for the betterment of everyone concerned. Tsander's diary reports a conference on 1 February 1932 convened by Tukhachevsky at the office of the Revolutionary Military Council in which commanders of the Air Force, Artillery Chemical Warfare, and others were present. Korolev and other leading GIRD members were also there as well as representatives of *Osoaviakhim* and the Leningrad GIRD (LenGIRD). The various justifica-tions for the proposed amalgamation were introduced and the machinery set in motion. Certainly increased funding (and wages) as well as facilities could be boosted immeasurably. However, the primary goals of the space-minded members would necessarily be suppressed in the interest of "national defense." A loss of independence and commraderie would also result. But these problems are not discussed by Soviet historians of rocketry and space travel.[105]

On 8 April 1932 a memorandum was sent by the heads of LenGIRD to Tukhachevsky reiterating their approval of the amalgamation plan. In the following month the General looked a quarter of a century ahead and noted in a report to the Chairman of the Council of Labor and Defense that the rocket principle offered unlimited possibilities for firing artillery projectiles of any power to any distance. In the field of aviation he foresaw "the ceiling of aircraft to the stratosphere . . ." The Russians have never admitted it, but clearly the military came first, and the dreams of flights to other planets were left behind. The pragmatist Korolev noted in his popular *Raketniy polet v stratosfere* (Rocket Flight in the Stratosphere) of 1934 that planes "must command the attention of all those interested in the field, rather than as yet unsubstantiated fancies about lunar flights and record speeds of non-existing airplanes."[106]

By August 1932, the Red Army's Department of Military Inventions was helping to fund MosGIRD but not LenGIRD. The merging process itself became bogged down in the mire of the bureaucracy and dragged out until late 1933. One of the problems was the mismanagement of MosGIRD. The Finance Department of the Red Army sent in auditors and found omissions in the bookkeeping but otherwise "the position with personnel is satisfactory." Korolev himself was called into account for the financial discrepancies. This, with the death of Tsander and the sluggish progress of the work naturally made spirits sag.

Both spirits and fortunes changed for the better, particularly with the launching of the "GIRD" 09 on 17 August 1933—the flight of the Soviet Union's first liquid-fuel rocket. Technically, it was a hybrid powered by liquid oxygen and solidified (jelly-like) gasoline. A true liquid-propellant rocket, the "GIRD-X", powered by liquid oxygen and alcohol, had been designed under the guidance of Tsander and was launched 25 November 1933.

A month earlier, however, on 31 October 1933, the Council of Labor and Defense at last adopted a resolution approving the Military Council's "Order on the Organization of the Reaction Propulsion Institute." This was the founding of the *Reaktivni Nauchno-Issledovatel'kii Institut* (Scientific Research Institute of Jet Propulsion, or RNII). It was hailed by the Soviets as the first state rocket research establishment in the world. A top-ranking military engineer and former chief of the GDL, Ivan Terentevich Kleimenov, was named head of the Institute with Korolev as his deputy. On 15 November 1933 the RNII was transferred to the People's Commissariat of Heavy Industry.[107]

In addition to Kleimenov, other GDL and LenGIRD members were also detailed to Moscow, along with equipment and projects. The premises of the new Institute on the outskirts of the city must have seemed palacial after GIRD's dank basement years. The building had been a diesel engine plant. Some of the engine beds remained and were soon modified to hold down rocket motors. A room on the Kolkhoznaya Square in Moscow was rented for the inauguration party of the RNII. Korolev appeared wearing for the first time a military tunic with the two pips signifying his rank of divisional engineer, or major general. He was 29 at the time.[108]

LenGIRD and RNII

From here we must retrogress to recount the story of LenGIRD. By comparison to MosGIRD (or GIRD), it is a brief but important one. LenGIRD began in the Spring of 1931 upon the initiative of a ship-building industry employee and recent graduate of the Naval Engineering School at Kronstadt, Vladimir Vasilyevich Razumov. Razumov's background in space travel theory came through reading Tsiolkovsky and through talks with Professor Rynin. At the same time, he was a consultant to a Leningrad group studying dirigibles and in this way became acquainted with the famous Russian aeronaut P. F. Fedoseyenko. The subject of interplanetary flight was brought up. To Razumov's surprise, the aeronaut quickly responded and proposed forming an organization of kindred souls to build rockets. Upon Fedoseyenko's suggestion, Razumov approached the Bureau of Air Technology of the Leningrad branch of *Osoviakhim*. Soon the popular scientific writer Yakov Perelman was also brought into the circle. Razumov recalled 30 years later that at the end of March, "I was with Rynin, discussing with him the calculation of rocket flight by a method which I had proposed, and casually mentioned to him that I had entered with the SPDACC *(Osoaviakhim)*, and would be working in an unofficial way on rockets with the Bureau of

Air Technology. Rynin responded with the fact that he also had spoken with the SPDACC on this question and that he, Gazhala and Perelman, had also joined together to participate in this work. Many times we gathered either at Rynin's or at my place to discuss at length, present propositions, and drew up plans regarding the development of work on rockets and the prospects of interplanetary communication. Thus was created the pioneer groups for studying jet propulsion at the Leningrad SPDACC comprised of Rynin, myself, Gazhala and Perelman."[109]

Responsibility for organizing the rocket group fell to Yevgeniy Chertovskiy, Vice-President of the Leningrad *Osoaviakhim.* The first general meeting was conducted 13 November 1931 in the Army and Navy House in which a presidium was selected and the LenGIRD officially instituted. Rynin, who was then completing his voluminous space encyclopedia, gave the opening address. Razumov followed with a report on the immediate possibilities of making some preliminary sounding rockets prior to exploring the cosmos. At the conclusion of the gathering, Razumov was proclaimed the President, and Perelman Vice-President. At the second assembly teams were set up similar to those of MosGIRD, and programs outlined. The five teams, each consisting of five to six individuals, included physicist Gazhala's scientific-research committee, and Razumov's Design construction sections. Perelman was placed in charge of "scientific propaganda." Another physicist, I. N. Samarin, headed the laboratory, and Chertovskiy was chief of the rocket range, a place yet to be selected. In fact, LenGIRD did not have a headquarters.

The first several meetings were held in members homes or the Army-Navy building. Chertovskiy solved this problem by getting permission to occupy a garret atop the Leningrad Engineering House. V. I. Shorin arranged for the rocketeers to be paid by *Osoaviakhim* for their materials. He also arranged instructive field trips to Moscow to the GIRD laboratory. Through Shorin's politicking, LenGIRD became affiliated with the Institute of Wire Communication, in whose well-equipped shops the first LenGIRD rockets were built. For additional money, Shorin arranged a trip for Razumov to Moscow to see the grand patron of Soviet rocketry, Mikhail Tukhachevsky. The General granted 15,000 rubles for the research.[110]

Like their MosGIRD cousins to the south, LenGIRD also gave courses and lectures. Here Professor Rynin was of great value as a teacher with his truly encyclopedic knowledge of the field. Mysteriously, he omits all the work of both LenGIRD and the older MosGIRD in *Mezhplanetnyie Soobshcheniya* (Interplanetary Flight), the last volume of which went to press in early 1932. Did the events happen too late to be included? Did he intend to save the details for his intended but never completed expanded issue of *Interplanetary Flight*? Or was he cautioned not to publicize any of the work which might have had military implications? We may never be sure of the answer. In his foreword to the ninth number of his encyclopedia dated 1 March 1932, he thanked GDL's Valentin Glushko and over 300 contributors from all over the Soviet Union. It is inconceivable for a man in his position and contacts not to have had more than an inkling of the activities in GIRD's Moscow basement or in the shops of the Institute of Wire Communications in his native Leningrad.[111]

Soviet historians are somewhat hazy about LenGIRD's final years but say it lasted up to the beginning of World War II. It continued its close cooperation with MosGIRD and in 1934 was transferred into a Reactive Motion Section headed by physicist M. V. Machinsky. This undoubtedly meant at least partial absorption into the RNII which had been just formed. In this respect it shared the same fate as the GDL. During the eight or nine year period of its existence, LenGIRD produced both solid and liquid-fuel rockets, some of the most interesting of which were equipped with photographic equipment, gyroscopes, and radio controls. For the Army, there were also new flare rockets and even message-carrying projectiles. Razumov never revealed LenGIRD's military connection, but it is clear from Tukhachevsky's early monetary gift, the nature of some of the projects, and the location of some of the tests, that the military was involved. LenGIRD also designed among the world's first rocket sounding vehicles. One was exhibited at the All-Union Conference on the Conquest of the Stratosphere held in Spring of 1934 at Leningrad's Academy of Sciences. Participation in this significant conference was one positive step towards genuine space studies. LenGIRD is also said to have investigated the effect of acceleration on animals, presumably like some of their contemporary American and German counterparts, working towards a better understanding of requirements for future life support systems and predicting man's biological behavior during space launches.[112]

Tsiolkovsky and the Rocket Societies

It was shortly before the All-Union Conference on the Conquest of the Stratosphere that Kleimenov, head of the RNII, opened correspondence with Russia's grand patriarch of space flight, Konstantin Tsiolkovsky. Kleimenov informed him of the amalgamation of the different rocket groups into the RNII in the name of "Communist reconstruction of all human society and for conquest of the heights of science and engineering." Kleimenov also beseeched Tsiolkovsky to join the RNII "as one of the three or four leading workers." Kleimenov obviously had an honorary and subordinate position in mind, but an immensely psychologically important one that was bound to add lustre and propaganda grist. The old man supported all of the rocket organizations of course, though himself had never actively experimented. His connection with the groups, however, could only be by proxy. This was dictated by his pastoral temperament as much as by old age. On the occasion of his 75th birthday in 1932, for example, Tsiolkovsky was invited to Moscow so that tributes could be paid. Golovanov describes his state of mind and why he never came: "When Tsiolkovsky was invited to Moscow he always refused, pleading indisposition, weakness, age, or deafness but, like Newton, he was simply a confirmed home-body. Any journey frightened him, the thought of hotels terrified him, and he recoiled from the idea of all frightful unaccustomed life when you don't know where and how you are going to eat or what you are going to sleep on. That was why the ceremonies held in Moscow and Leningrad to mark Tsiolkovsky's 75th birthday were celebrated without him; despite all efforts to persuade him to attend, Tsiolkovsky stayed at home. But at the end of November he had to come and in the Kremlin M. I. Kalinin invested him with the Order of the Red Banner of Labour."[113]

Tsiolkovsky communicated with the RNII until his death in 1935, exchanging technical reports and notes. Some were posthumously published in the serialized RNII papers *Raketnaya tekhnika* (Rocket Technology), which first appeared in 1936. In addition to this honor, Tsiolkovsky was elected an honorary member to RNII's Engineering Board in March 1934. Ironically, because he rarely left Kaluga, Tsiolkovsky probably never witnessed a rocket shoot of either GIRD, LenGIRD, or RNII.[114]

The Purges

If the earliest years of the Soviet rocket societies are somewhat unclear, then the latter phase, just preceding the war, is shadowy at best. Soviet sources are reluctant to admit difficulties. Glushko, in his *Ragvitiv Raketostroyenia i Kosmonavtiki v. SSSR* (Development of Rocketry and Space Technology in the USSR), reveals that matters were not always smooth. In 1935, L. K. Korneyev, Tsander's biographer, set up Design Bureau Seven (KB-7) as part of the RNII. Writes Glushko, "It failed to produce the desired results and was dissolved." The true reason may have been the terrible purges of 1937—1938. They were so devastating that they seem to have resulted almost in the dissolution of the RNII itself.[115]

The Soviet rocketry chronicler, and participant I. A. Merkulov, dates the termination of the RNII to 1938. Other accounts are inconsistent and say it thrived after this time. Merkulov also indicates that from the start it came under the thumbs of the military, first functioning "within the system of the Military-Science Committee (of the Central Council of *Osoviakhim*), and then as part of the Stratosphere Committee." In any event, the military through Tukhachevsky always controlled the direction of much of the work.

We can conjecture from the available evidence that with Stalin's liquidation of Marshal Tukachevsky in 1937, on the baseless charges of "passing military intelligence to a foreign power" (i.e., Germany), sabotaging the Red Army, and promoting the restoration of capitalism in the USSR, that the reverberations directly affected his subordinates within the RNII. We know of the deaths of several leading rocketeers during this period: Chief of the RNII (I. T. Kleimenov, 1898—1938); Deputy Chief of the RNII Georgii Erickhovich Langemak (1898—1938); and Chief of the GDL, Nikolai Yakovlevich Il'yin (1901—1937). Langemak's fate is confirmed by the euphemistic admission of Slukhai, who says he was "slandered" in 1937. How many others perished in the purges cannot be determined. Before the amalgamation, the old MosGIRD organization and its affiliates around the country numbered about 1,000 persons. According to Glushko, LenGIRD amounted to more than 400 by 1932. Figures for the GDL and RNII are unavailable. Very probably the aggregate of the RNII-GDL of the 1937—1938 period remained around 1,000 more or less. By 1938 the RNII had apparently been swept away with no reason stated in the histories. G. A. Tokaty, a post-war Soviet rocket scientist who later defected to the

West, affords us another viewpoint. Addressing the question of a Soviet lag in rocketry during World War II vis-à-vis the Germans, Tokaty said: "I am sure that the historians will find this question exceedingly interesting; but there is no simple answer. We can, however, recall one or two major facts which, no doubt, contributed to our failure. I think that the (now officially acknowledged) political arrests and murders of the 1935—40 period, known as the *Ezhovschina* (*Ezhovism,* after Ezhov, then the NKVD secret police chief) caused much greater damage than is realized abroad. Far too many scientists, technologists and managers were destroyed, humiliated, or disheartened. And rocket experts were no exception. Then as I have already said, Marshal Tukhachevsky was the top governmental spiritual leader of military rocketry. But, thanks to Nazi provocation, he was shot in 1937 as a 'German spy,' and this sparked off a whole chain of disasters. Almost all who worked on a project discussed with or authorized by him, or who were in contract with him—as all leading rocket specialists were—had now to face the danger of being proclaimed an 'accomplice of a spy.'"

It is a curious fact that an often published Soviet photograph of the GIRD-X rocket launch crew identifies only one of the 12 persons present. This is Sergei Korolev. The unidentified individuals may have been purge victims. Korolev himself was arrested in 1937 "along with all the others" of his organization according to Oberg, but was later allowed to continue his rocket work.[116]

Shchetinkov, a former member of the RNII, implies in his 1969 paper, "Main Lines and Technical Research at the Jet Propulsion Research Institute (RNII), 1933—1942," that the organization continued until some years later than the purges. However, he discusses no projects for 1939 and for the years 1940—1942 covers only the RP-318 rocket fighter program and its follow-up, the BI-1. Further, Valentin P. Glushko's original GDL team functioned as a sub-unit of the RNII from 1933 to 1938, and was then begun again in 1941 as the GDL-OKB (Gas Dynamics Lab—Special Design Office). The history of this particular unit and of Glushko's career is of special importance as the GDL-OKB was not only the sole surviving rocket organization from the purge days but continued to exist and later produced rocket engines for the first *Sputnik* and manned spaceflight launches. Glushko and the essentially military GDL-OKB thus represents a direct link from the RNII to the Space Age.

Following Glushko's chronology in his *Rakenyye dvigateli* GDL-OKB (Rocket Engines of the Gas Dynamics Laboratory Experimental Design Bureau), in 1939 the GDL sub-unit "separated from RNII and was made into an independent design group at the Moscow Aviation Engine Plant." Glushko was still in charge. In a memoir article in *Aviatsia i Kosmonavtika* for October 1974, Glushko recalls that also in 1939 he had "a bit of luck" in acquiring Professor G. S. Zhiritskiy, a specialist on excavating machines, as his assistant who later developed turbines and centrifugal pumps for large liquid engines. In the Fall of 1940 the group moved to Kazan where they continued their work, mainly developing technical documentation for the GG-3 gas generator with automatic feed from a turbo-pump unit and a project for the installation of auxiliary (Jet-Assisted Take-Off) rockets for the S-100 aircraft. The GDL-OKB was then formed the following year and at Glushko's "solicitation," Korolev became Glushko's deputy for flight testing though Zhiritskiy was called the "Deputy Chief Designer." Here it is interesting to note Oberg's remark that "Glushko, an old friend and colleague of Korolev's from the GIRD days . . . was somehow still a free man."

Besides Korolev and Gluskho, several other former GIRD and RNII members miraculously emerged from the purges unscathed and continued to work to become the rocket builders of the *Sputnik* era. Amongst these were Tikhonravov, Dushkin, Merkulov, Pobedonostev, and Rauschenbakh. There is a striking parallel between the story of these Soviet pioneers and those of the old German Rocket Society, the VfR. In both countries the men started in the early 1930s as idealists pursuing rocketry as an avocation with a vision towards an eventual space or stratospheric vehicle. In both countries one of their number was to emerge as exceptional leaders—von Braun in Germany, Korolev in the USSR. By 1935 the Armies of both nations clearly diverted the idealistic path towards militaristic aims. At the opening of the Space Age, von Braun directed the United States' Explorer 1 mission and afterwards the Apollo program. In the Soviet Union Korolev had become the "Chief Spacecraft Designer" who directed the construction and launch of the first *Sputniks* and manned *Vostoks.* In both countries the respective design and launch teams also included several other 1930s rocketeers. The legacies of the VfR, GIRD, RNII and GDL-OKB is manifest. They served as experimental test beds and schools for far greater things to come.[117]

Soviet Technical Developments 1930s

Of the Soviet hardware of the period a great deal has been written. The accomplishments were impressive. Merkulov writes that between 1932 and 1941 more than 100 rocket engines of various types were designed in the USSR. Though many of these projects were military in nature and not scientific or space-oriented. The following is a general survey of strictly experimental or stratospheric sounding rockets. A table of Soviet scientific vehicles that flew is found in the Appendix.

The historic launches of the GIRD 09 and GIRD-X rockets, the first hybrid and liquid vehicles, respectively, in the Soviet Union have already been mentioned. Other salient characteristics of these important flights are that the 09 rocket, launched 17 August 1933, went to 400 meters (1,312 ft) with the liquid oxygen/solidified gasoline 09 motor producing 52 kilograms thrust (114.6 lbs) for 15—18 seconds. Somewhat later, a modified version reached about 1,370 meters (4,500 ft). The fuel, sometimes called "solidified benzine," originated thousands of kilometers away in the oilfields of Baku. It was a solution of colophony, a natural resin in benzine. Rather than a solidified stick of this substance in the manner of Hermann Oberth's hybrid *Kegelduse* motor, the viscous fuel was fed into the combustion chamber through a grid arrangement and met the incoming oxygen that was fed in by its own vapor pressure. An aircraft spark plug ignited the mixture.

Problems and delays encountered prior to the launch of the 09 were as vexatious and frustrating as those met in count-downs of the early Space Age. One delay was caused by foul weather. The GIRD truck had also overturned in wet snow on the way back. Another time the truck nearly overturned again. It had struck a loose cobblestone in the road to the Nakhabino launch site outside Moscow but the GIRDers were alert and snatched up their valuable cargo in time to cushion it from the jolt. Finally, when 09 did go off at five o'clock in the afternoon, Korolev was incredulous. "She really flew," he said almost in disbelief. "She really flew. We didn't do all that work for nothing." Korolev's official report was less animated. It also spoke of a successful flight but with a tumbling crash back to earth due to a lateral thrust caused by the burn-through of a flange.[118]

There are fewer recorded anecdotes about GIRD-X but the launch on 25 November 1933 was likewise not smooth. The fundamental design was worked out by Tsander. The liquid-oxygen-ethyl alcohol engine developed a thrust of 70 kgs (154 lbs) for 22 seconds (some references say 12—13 seconds). The flight was brief. The 29.5 kg (65 lbs) 2.2 meter (7.2 ft) long rocket went to 75—80 meters (246—262 ft) vertically, then veered abruptly and fell at a distance of 150 meters (492 ft). As a true liquid-propellant vehicle, the GIRD-X nonetheless represents a greater milestone than the 09 in that it served as the basis for more advanced designs developed between 1935 and 1937. This also made it the direct forerunner of the *Sputnik* rockets.[119]

Other important Soviet experimental, non-military projects may be divided into three categories: so-called "winged-rockets," sounding rockets, and ramjets.

The phrase "winged-rockets" is a now antiquated Soviet phrase of the 1930s. To Western ears it is a superfluous term and means nothing more than any rocket, regardless of its mission, that is fitted with wings or fins. The Soviets had two definitions. One is the manned or unmanned rocket plane. The second meaning is a winged, unmanned rocket-propelled missile of any size. Langemak and Glushko, staff members of the military GDL organization, provide a good contemporary description of the latter, stressing the military implications: "The winged rockets, sometimes called aerial torpedoes, have a special place among rockets. They have, so to speak, an intermediated position between artillery shells and aircraft. They are the same as the former in their operational employment and high rate of travel, but in their appearance the aerial torpedoes resemble airplanes, for most of them possess wings and an automatically operated controlling device [i.e., automatic guidance]."[120]

It is not always clear in Soviet discussions of "winged-rockets" of the 1930s which of the definitions is meant, in short, whether the subject is an airplane or missile. Matters are further complicated in that small aerodynamic and structural test models of the planes were also built. One thing is certain: the military dominated all work on aircraft and therefore controlled much of the stratospheric research. Sergei P. Korolev who championed the rocket plane as a logical means of exploratory flights leading to true spaceships was compelled to pay considerable attention to the military implications of this type of rocket and to downplay its interplanetary role. In his *Raketnyi polyot stratosfere* (Rocket Flight in the Stratosphere), published by the *Voyenizdat*

(Military Press) in 1934, he urged the development of "winged rocket" weapons, or guided missiles, especially as he believed the West was progressing along these lines: "It is understandable that in the imperialist countries the rocket will be used least of all for scientific and investigative purposes. Its chief assignment will be for war, for the significant altitude and distance of its flight are the most valued qualities for this use."

Korolev later designed not only missiles but also rocket-fighter interceptors and JATO's for military planes. At the same time, he never lost sight of building a peaceful high-altitude rocket-powered flying laboratory. He espoused the "scientific and investigative" rocket plane in his important paper "A Winged Rocket for Manned Flight", presented at the First-Union Conference on the Use of Reaction Aircraft for Conquering the Stratosphere, held in Moscow, 2—3 March 1934. The idea of stratospheric rather than interplanetary rocket craft is also expounded in later papers. Privately, he confided to Yakov Perelman, the famous space travel publicist in a letter of 18 April 1935, his real reasons for downplaying the interplanetary aspect of rocketry: "I only wish that in your future works (since you are a specialist in rocketry and an author of a number of excellent books), you would pay more attention not to interplanetary problems, but to the rocket engine itself, to a stratospheric rocket, etc., since all this is now closer to us, more understandable, and more necessary. A great deal of nonsense has been written on interplanetary themes, and it is now hurting us badly. Just a few days ago, they said directly to me in one journal: 'We are avoiding publishing material on rocketry, because they are all lunar fantasies, etc.' Now I have even more difficulty in convincing people that this is not so, that rockets are a defense and science."[121]

As to specific technical developments with "winged rockets," there is ample evidence to support the above contention of military domination of the research. It may be summed up briefly.

Korolev's "winged rocket" plans led to the formation of MosGIRD in 1931 with the object of adapting Fridrikh Tsander's OR-1 engine to an aircraft. This was never accomplished. The so-called RP-1 (Raketoplan-1) was built but not flown under rocket power. Various smaller rocket planes were also constructed, primarily after GIRD and GDL amalgamated in 1933. Ostensibly these vehicles served as aerodynamic and structural test models. But as Shchetinkov states, "Over the period of 1933—39, RNII also developed winged liquid- and solid-propellant rockets. The main purpose here was to investigate the flight dynamics of winged jet [i.e., rocket] vehicles . . . and the practical application of these rockets in air defense and ordnance systems . . ." The 212 winged rocket is the most well known project of this category. Designed by Korolev in 1936, the 212 was a gyro-stabilized long-range anti-aircraft missile with a calculated range of 50—80 km (30—50 mi) and propelled by the highly reliable ORM-65 nitric acid engine of 150 kgs (330.6 lbs) thrust. Of the launch weight of 210 kgs (462.9 lbs), according to Astashenkov, 30 kgs (66.1 lbs) constituted the "warhead." Numerous bench tests of the weapon were made up to 1939, but the project was never completed, probably because of the war. Astashenkov also mentions two other winged rockets: "Rocket 201 . . . might be put into the 'air-ground' [i.e., aircraft] class. Rocket 217 might be called an anti-aircraft rocket, with aiming along a projected beam. Incidentally, the group which Sergei Pavlovich [Korolev] headed was planning to install self-steering equipment on its winged rockets." The latter weapon which came under Korolev's jurisdiction, was solid-propelled and was completed by 1939 at the RNII. Korolev's other winged rockets during this period culminated in the RP-318, based upon his earlier SP-9 sailplane. It flew successfully as the Soviet Union's first manned rocket plane on 28 February 1940. But again, the end result was not a research vehicle but a military machine. Repeated tests of the RP-318 led to the BI-1, a rocket fighter.[122]

Soviet sounding rockets of the 1930s were more bonafide scientific tools in the peaceful exploration of space or near space than were the so-called "winged rockets." Alexsandr Ivanovich Polyarny of the Scientific Institute of the Civil Air Force designed the Soviet Union's earliest known meteorological sounding rocket in 1931. It was a modest solid-propellant vehicle carrying a radio telemetry package up to a planned altitude of 6,000 meters (19,686 ft). High energy solid propellant studies were made but the rocket project was not realized as Polyarny was transferred to the Institute of Aircraft Engine Construction; at the same time he was invited to join GIRD and thereafter worked closely with Tsander.

In Leningrad, in 1932, V. V. Razumov of LenGIRD also proposed a high altitude solid-propellant (smokeless pyroxoylin) rocket, but as it was designed to carry up a camera to 10,000 meters (32,810 ft) "for taking photographs of terrain" probably military application (i.e., reconnaissance) was intended. Razumov also proposed a "recording rocket" with an ascent altitude of 10 km (6.2 miles) "for obtaining data on barometric pressure, temperature, and density of atmospheric air at heights of 0—10 km above sea level." These plans were too ambitious for the state-of-the-art but scaled-down versions were produced and flown up to one kilometer (3,280 ft) at the Leningrad *Osoaviakhim* camps under the direction of LenGIRD member M. V. Gazhala. It is still difficult to pin down the first bonafide Soviet sounding rockets as the "M. V. Gazhala experimental rockets," as they were also called, came in "high altitude, [propaganda] leaflet [dispersing] and shrapnel simulation variations." The overall length of these rockets was 2.1 meters (6.8 ft), the diameter 0.23 meters (.75 ft), the fully loaded weight 30 kgs (66.1 lbs) including 10 kgs (22 lbs) of propellant and a 5 kg (11 lb) payload package, presumably with parachute. The original Razumov 10-km design was a nose-drive configuration with four aft fins providing stability. It is assumed this same configuration was chosen for the scaled-down versions. One interesting feature of the design was that cooling of the motors was affected by annular channels within the body that permitted air to flow over the combustion chambers.

In 1933 when LenGIRD switched to liquid propellants, their first project was the Razumov-Shtern Recording Rocket. This implied a scientific sounding mission. The most unusual feature of this vehicle was the "rotary rocket engine" designed by LenGIRD member Alexandr Nikolayevich Shtern. Shtern sought to solve two problems faced by rocket engineers of the day: providing a constant propellant flow to the engine and insuring overall stability. His solution was to place two rocket chambers opposite each other and mounted to a horizontal fuel pipe. The pipe was attached to a bearing leading from the lox-gasoline propellant tanks above; within the bearing was a valve which initiated the propellant flow. The nozzles of the chambers were canted so that both chambers would spin on the bearing. Shtern thereby theorized that the propellant could be fed to the chambers by centrifugal force. At the same time the spinning chambers were supposed to cause a gyroscopic effect which would stabilize the rocket. He calculated that the engine could deliver 200 kgs (440.9 lbs) thrust for an exhaust velocity of 2,000 m/sec (6,560 ft/sec). The maximum estimated velocity of the 90 kg (198 lb) rocket was 100 m/sec (328 ft/sec) thus sending the rocket up to 5 km (3.1 miles).

According to Razumov, the rocket body, combustion chamber and nozzle were built in the shops of the Leningrad Institute of Wire Communication and displayed along with a design drawing at the All-Union Conference on the Conquest of the Stratosphere held in Leningrad from 31 March through 6 April 1934. Perhaps only one of the canted nozzles and chambers were made. The full engine was still not ready when LenGIRD decided to make preliminary aerodynamic checks on the rocket body. In place of the Shtern liquid propellant rotary unit, a solid-fuel powder engine designed by V. A. Artem'yev was installed with a successful launch made at the end of 1934 at the Aerolitic Institute Station in Slutsk. Soviet rocketry historians offer no performance details on the flight. Tikhonravov and Zaytsev says only that the building of the original liquid engine "dragged on until March 1935, when serious complications caused this effort to be abandoned."

In effect, the aborted Razumov-Shtern Recording Rocket was LenGIRD's last project. In a memoir paper Razumov says that on his own initiative and in consultation with Rynin, he designed two other liquid fuel sounding rockets, one planned for altitudes up to 300 km (186 miles). However, he continues, the "first stage of rough planning" terminated on 25 November 1933 and further development was "prevented by unforeseen circumstances." Perhaps this remark refers to the breakup of LenGIRD and its absorption into the RNII on 31 October 1933.[123]

The RNII also attempted to develp sounding or "stratospheric rockets." In this they were no more successful than LenGIRD. During 1933 and 1934 V. S. Zuyev, a former GIRD engineer, designed and built a 50 km (31 mile) altitude cigar-shaped vehicle but was not able to produce the engine. Tikhonravov and Zaytsev say that the rocket's components (presumably the empty body, nosecone and four fixed fins) were later incorporated into another high-altitude rocket. Later on, they continue, a rocket of the Zuyev design was fitted with an 02 engine—possibly the 100 kg (220 lb) thrust alcohol-oxygen OR-2 or 02 engine of the RNII, mid-1930s—and made flights. Presumably the performance was considerably more modest than Zuyev's original goal of 51 km as the latter figure would have been a then remarkable world's record.

The *Osoaviakhim,* Design Office No. 7 (KB-7), and the All-Union Aviation, Scientific, Engineering and Technical Society also designed, built and flew sounding rockets. In all cases, RNII material or personnel were used. A. I. Polyarny, who had begun working for the *Osoaviakhim* in the Autumn of 1934, but still apparently attached to the RNII, collaborated with E. P. Sheptitskiy on the development of a liquid oxygen/ethyl alcohol meteorological rocket. This was a very simple but reliable rocket, the fuel being fed into the combustion chamber by partial pressurization of compressed air and the oxidizer by its own evaporation. This engine was designated the Polyarny M-9. It produced a maximum of 40 kgs (88 lbs) thrust for 14 seconds. Total propellant weight was 2.4 kgs (5.2 lbs), payload weight 0.5 kgs (1 lb), and the rocket's total weight, 10 kgs (22 lbs). The *Osoaviakhim* or *Osoaviakhim R-1* rocket as it was officially designated, stood about 1700 millimeters (66.9 in) high, including the long fin assembly which comprised about two-thirds of the rocket. The diameter was 126 mm (4.9 in). Estimated altitude capability was 5,000 meters (16,405 ft). By the Spring of 1935, the rocket had been built and successful bench test runs made. In his own account of the *Osoaviakhim R-1,* Polyarny says nothing of any attempted launches, though Tikhonrarov and Zaytsev write that a flight experiment made at that time at Nakhabino, near Moscow, failed. Apparently both the ignition and propellant valves were activated manually. At launch time these could not be properly coordinated and a special key that opened the tank valves was not removed quickly enough. Ignition had already taken place. The rocket went up but wedged in the launcher—probably because of the protruding key that should have been removed. The engine burned until the propellant was consumed but the rocket was prevented from properly flying. The *Osoaviakhim R-1* rocket still showed promise.

Polyarny was able to salvage the project when he began working as deputy director of another Soviet rocket organization, Design Bureau No. 7, or KB-7, created in the Summer of 1935. The nature of this organization is not clear. Since its rocket test station was designed by the Kuibyshev Military Engineering Academy, it is likely that it was military. Tikhonravov and Zaystsev say that the *Osoaviakhim* rocket designed by Polyarny was redesignated the R-06 when turned over to KB-7 and that "it had a different role to play." The role is not spelled out. Dimensions of the KB-7 version of the R-06 did not differ from the original rocket. Improvements were mainly in the launching and ignition system. The R-06 also underwent wind-tunnel tests in the Central Institute of Aerohydrodynamics (TSAGI).

Nine R-06 rockets were built and launched between early 1937 and early 1938. The flight missions were not mentioned but may have been ballistic studies (i.e., determining bombardment ranges) as the launchings were made at different angles to the horizon. Flight stability was dependent upon wind velocity and direction. Maximum range, not altitude, is stated as being about 5 km (3 miles). The average altitude therefore amounted to about half that figure which was hardly practical for a potential sounding vehicle. Instability was now the major problem, but this was rectified with the installation of an in-board gyroscope. This change necessitated a modification of the stabilizers. Dimensions still remained the same except that the height of the basic rocket increased from 1,700 mm (66.9 in) to 1,285 mm (50.5 in). The rocket also received a new name: ANIR-5. Polyarny says that six ANIR-s's were built and flight tested but again, a minimum of details are offered. Perhaps the dissolution of KB-7 about 1939 was also responsible for the abandonment of ANIR-5.[124]

If the R-06 and ANIR-5 rockets were not bonafide sounding vehicles, KB-7's R-05 rocket was. The R-05 was an ambitious project with an interesting history though the full vehicle was never flown. Polyarny and P. I. Ivanov were the principal designers, KB-7 receiving the assignment for the rocket in late 1937. The requirements for the project emanated from outside KB-7, namely the Geophysical institute of the Academy of Sciences of the USSR. O. Yu. Schmidt, Director of the Academy, desired a rocket capable of a 50-km (31 mile) altitude for space radiation studies. The Leningrad Institute of Physics also showed keen interest and support of the R-05. Other scientific organizations also later afforded support. Some Soviet writers consider the R-05 to have been a two-stage rocket. Since the two solid-propellant boosters mounted at the base were to burn simultaneously with the main liquid or sustainer engine and then jettisoned, the vehicle may properly be called a stage-and-a-half rocket. The rocket stood 2,250 mm (88.5 in) high from nosecone tip to sustainer nozzle base. The diameter was 200 mm (7.8 in). Launch weight, with boosters, was 60.5 kgs (133.4 lbs), according to Tikhonravov and Zaytsev; Merkulov puts the booster assembly as 14 kgs (30.8 lbs), and the sustainer at 55 kgs (121.2 lbs), thus totaling 69 kgs (152.1 lbs). Average thrust of the boosters was calculated at 200 kgs (440.9 lbs) each, according to Tikhonravov and Zaystev, while Merkulov records 1,500 kgs (3,306.9 lbs) maximum for the combined boosters. Burning time for the boosters was 2.5 seconds. Propellant weights amounted to 6 kgs (13.3 lbs) for the boosters and 30 kgs (66.1 lbs) for the 96% ethyl alcohol/liquid oxygen sustainer.

The payload, with parachute was 4 kgs (8.8 lbs). RNII engineer F. L. Yakitis developed the M-29e rocket engine which featured a regeneratively-cooled nozzle and a solid-propellant hot-gas generator with an operation time of 40—42 seconds. The designed thrust rating of the M-29e was about 175 kgs (385.8 lbs) for 25 seconds. (Polyarny says 37 seconds). The R-05 was also to be fitted with a gyroscopic stabilization system designed by P. I. Ivanov who had worked out a similar system for the R-06. Steering of the R-05 was to be partly accomplished by an infrared beam-riding system which was worked out in 1938 by the Ukranian Physiotechnical Institute. A narrow infrared beam was projected in the target direction and reacted upon photo-electric cells mounted on the rocket's stabilizers. This system was to only operate up to 10 km (6.2 miles) of the rocket's course which was sufficient because this distance was about the length of the powered portion of the flight. Also, due to the inverse square law, the beam intensity would decrease in inverse proportion to the square of the distance (if the distance were doubled, the beam strength would be one-fourth).

By the Summer of 1938, six scaled-down models of the R-05 were built and successfully flown in order to test the gyroscope. These rockets were not powered by Yakitis engines, but smaller, unspecified alcohol/liquid oxygen units. Meanwhile, the Yakitis M-29e was developed, static tests showing good results with the hot-gas generator. At the same time a great deal of attention was paid to the R-05 payload. Among the several instrument packages was a miniature camera for automatically photographing the Earth's surface at different intervals throughout the flight, barometers, a noninertial thermometer, an accelerometer, a gauge for recording engine pressure, and a radio transmitter for relaying back the data in code. The camera was designed and manufactured by the Leningrad Optical Institute on KB-7 specifications. The radio transmitter-coding device which was linked to the other instruments was especially developed and made by the Main Geophysical Observatory of the Hydrometeorological Service. In addition, the P. K. Shternberg Astronomical Institute devised a multi-camera range-finder system in which the cameras were stationed at key points in a geodesic layout. A luminous, pyrotechnic trail on the rocket made for easier photographic detection.

For all the technical input and thousands of rubles expended on the R-05 the project never reached fruition because of "organizational complications," according to Tikhonravov and Zaytsev. Undoubtedly, there were some technical problems too, though it is interesting to note Glushko's corrobrating statement that "Design Bureau Seven (KB-7), as it came to be called, failed to produce the desired results and was dissolved about 1938 or 1939." Yet this was still not the demise of the project. According to Tikhonravov and Zaytsev: "At the beginning of 1940, the Moscow Higher Technical School (MVTU), supporting the initiative of the rocket's developers, agreed to complete the R-05 rocket on the condition that a customer be found. The Hydrometeorological Service of the Council of People's Commissars or the USSR soon agreed to finance this work, but the war terminated these efforts a few months later." It is also interesting to note, as mentioned by Polyarny, that a variant of the R-05, known as the R-05g, was also designed by KB-7. As he says this rocket was to fly "at an angle to the horizon," there is the possibility that like the R-06 it may have been also planned as a bombardment projectile.[125]

The bonafide sounding rocket of the All-Union Aviation, Scientific, Engineering and Technical Society (AVIAVNITO) had a moderate success. Tikhonravov and Zaytsev say that in 1935 there were some "unused components" at the RNII. RNII personnel decided to use these to make a "stratospheric" rocket. With a 5,000 ruble grant from AVIAVNITO the work proceeded the following year. One of the components was a powerful ceramic aluminum oxide and magnesium oxide lined 12-K 95—96% alcohol/liquid oxygen engine, designed by Leonid S. Dushkin, that produced 300 kgs (660 lbs) thrust for 60 seconds. Tikhonravov and Zaytsev provide no background on this engine but Dushkin says it was "earmarked for use in wingless and winged rockets," possibly military missiles. It was well suited for its new high altitude task as it weighed only 15.5 kgs (34 lbs). The overall weight of the rocket, called the *Aviavnito,* was 97 kgs (213.8 lbs), according to Merkulov. Of this the propellants weighed 32.6 kg (71.8 lbs). The rocket was also kept light because of V. S. Zuyev's contoured and hollow stabilizer wings. The payload amounted to 8 kgs (17.6 lbs) and consisted of a parachute and a simplified baragraph designed by S. A. Pivovarov for measuring the altitude. The overall length of the *Aviavnito* was 3,155 mm (124.2 in), the diameter 300 mm (11.8 in).

There is a wide discrepancy as to the design altitude of the *Aviavnito*. Tikhonravov and Zaytsev say it was 3,800 meters (12,467 ft) while elsewhere Tikhonravov says 10,800 meters (35,433 ft). Dushkin also says 10,800 meters. Real performance matched the former figure. The first flight took place on 6 April 1936 and received press coverage in *Pravda,* though performance values were not released at that time. The reporter presented only an overstated narrative: "The engineer has switched on the electric ignition plug. Gray smoke of evaporating propellant. Spark. And suddenly a dazzling yellow flame appeared at the base of the rocket. The rocket moved slowly up the guide rods of the launching frame, slipped out of its steel embrace and rushed upwards. The flight was an extremely impressive and beautiful spectacle. A flame flew from the motor's nozzle, and the gas efflux was accompanied by a deep hollow roar. After reaching a certain altitude, the white parachute opened automatically and the machine descended slowly to the snow-covered field."

The launcher for the initial flight was modified from one made by Mikhail K. Tikhonravov for the GIRD 07 rocket which had not been flown. A photograph of this launcher shows it to have been an excessive arrangement of pipes as the GIRD 07 was a complex delta configuration. Probably for this reason a newer, simpler launcher was made for successive flights of *Aviavnito*. This consisted of a wooden mast similar to a standard Soviet 48-meter (157.4 ft), radio tower, but with a launch rail made from a length of narrow-gauge railway rail attached to it. Lugs on the side of the rocket slid over the rail. This launcher was first used in the second flight attempt of the *Aviavnito* on 2 August 1937. The maximum pressure gauge in the propellant tanks failed and the launching was aborted. On 15 August the take-off and flight were successful, though after it opened the parachute became detached from the rocket and the rocket was obliterated upon impact. The baragraph was recovered and was found to read 2,400 meters, (7,874 ft). RNII and *Aviavnito* engineers assumed this was the altitude at which the parachute opened and therefore credited the rocket with reaching more than 3,000 meters (9,843 ft). Dushkin calls this flight "more successful" than the first attempt so that we can conclude the initial attempt was well below 3,000 meters. Merkulov's list of Soviet rockets indicates that two *Aviavnito's* were built but no information is available as to whether any additional flights were made.[126]

Finally, three other Soviet sounding rockets of the 1930s should be briefly mentioned, although they never achieved flight.

From 1936 to 1937 a meteorological rocket was built in the workshops of the Scientific Research Institute of the Civil Air Fleet on funds of 5,000 rubles and based upon the design of A. F. Nistratov and I. A. Merkulov, two ramjet pioneers. Tikhonravov and Zaytsev say that this water-cooled rocket was built but never tested.

Dushkin and Tikhonravov collaborated on another sounding rocket in 1937, a 30 km (18.6 mile) vehicle. It too was not completed but scaled-down versions were flown with powder rockets which were literally relics of another era; they had been in Army storage since 1916! Probably these were signal or flare rockets, propelled by gunpowder. They sufficed to test the stability of the planned configuration. Other models were built for the Dushkin-Tikhonravov project including a unique torus, or doughnut-shaped liquid propellant tank arrangement. The torus tanks were tested for strength by pressurizing them but the rocket itself was not built.

Dushkin conceived another stratospheric rocket in 1937 which was built and was to have been fitted with an improved 12-K engine designated the 205. Like the 12-K, it was ceramic-lined and produced a thrust of 150 kgs (330 lbs). The 205 was apparently designed by Valentin P. Glushko for use on some of Sergei P. Korolev's "winged rockets." The Dushkin rocket was built, along with the 205 engine and four automatic gryoscopically-control systems for the rudders. Yet, according to Tikhonravov and Zaytesev, "Since this stratospheric rocket was designed for vertical flight and did not carry out tactical [i.e., military tasks], further finances were unavailable, and it was not produced."[126]

Ramjets

GIRD and RNII also experimented with ramjets. Ramjets are air-breathing engines operating at supersonic flight. Essentially, the ramjet is a specially shaped tube which is literally rammed into the air by a rocket booster until operating speed is attained. Fuel is then injected and burned with the air which is built up and compressed at the forward end of the tube by a diffuser. There is no moving compressor as in a conventional jet engine. The exhaust gases are expelled from the rear of the tube.

Because ramjets can only operate in the air, their use does not seem to include spaceflight. The GIRD experimenters and Soviet historians justify ramjets for space flight, however, by pointing out that both Tsiolkovsky and Tsander thought of them as boosters in the atmospheric portion of a rocket flight. Yuri A. Pobedonostev, recalling his own contributions as the first GIRD ramjet team chief, stresses that "all investigations and design by ramjet engines were performed by space technology enthusiasts of the [*Osoaviachim* Central-Council] Stratospheric Committee without compensation." In all probability the joint GIRD-Osoaviakhim ramjet investigations, placed in charge of Pobedonostev in 1931, began as exploratory projects. But ramjets can be turned into weapons. *Osoaviachim* was also an acronym that stood for Society for Assisting Defense and Aviation and Chemical Construction in the USSR. One later Soviet ramjet vehicle was in fact referred to as a "wingless torpedo." The ultimate purpose or trend of Soviet ramjet work is therefore not always clear though the end result of some of the work was the attempted boosting of speed of aircraft. In any event, overall results were negligible. Soviet ramjet pioneers found these engines consumed prodigious amounts of fuel. Thrusts were also offset by drag. Because of the uncertainty of the direction of early Soviet ramjet research and its later military rather than space flight application, this topic is not considered further.[127]

Soviet Contributions: An Overview

Ironically, despite enormous technological strides made by GIRD, RNII and other agencies, a primitive but effective solid-fuel barrage weapon called the *Katyusha* was the only Russian rocket of any consequence used during World War II. What had happened?

The purges are part of the answer. Besides this, Soviet rocket projects of the 1920s and 30s, whether for peace or war, were experimental with none ready for mass production by the opening of hostilities except the *Katyusha* class. The Soviets had fragmented their various rocket projects with the majority incompleted. On the eve of the German invasion of June 1941, Soviet rocket technologists who had survived the purges were consequently involved with relatively inconsequential projects. Korolev and Glushko, for example, together concentrated on liquid-propellant rocket-assisted take-off units for aircraft, though efforts were also made towards developing bombardment missiles and the first rocket planes. This work was carried out by the GDL-OKB which went on to produce the hardware for the *Sputniks.*

The native talent—principally Korolev, Glushko, Dushkin, and others—were responsible for picking up the momentum after the war. Yet there were other inputs into modern Soviet rocketry which are pertinent to consider briefly. These were German V-2's and technicians captured by the Soviets. How much was learned and did it cancel out what was "lost" during the purges and the war years?

G. A. Tokaty, the man in charge of rounding up German rocket scientists for the Russians, believes the Soviets were "as advanced, as inventive, and as clever as the German rocketists [sic]. But in putting these theories into practical technology, we appeared to be miles behind the Germans." In comparison to the V-2 technicians gained by the West, Tokaty also says the quality of V-2's and German rocket scientists obtained by the Soviet Union were less than optimum compared with those acquired by the Americans. He thereby concludes that the Soviet Union gained very little from V-2 technology and that subsequent Russian successes in rocketry and space travel were largely due to native talent.

Yuri Pobedonostev, a GIRD veteran of the 1930s, reached the same conclusion in a January 1960 article in *Astronautics*. Pobedonostev also says that according to Wernher von Braun, the Russians actually made poor use of the German scientists and that "the Germans were really isolated from the real Soviet rocket program." Ordway and Sharpe, historians of the von Braun team and the V-2 rocket program, likewise conclude that the Russians learned little and "deprived themselves of many valuable and creative contributions that these men could have made to Soviet rocketry." The ultimate conclusion is that the GIRD and RNII experiences did mean something. Through perseverance, and under sometimes extremely difficult circumstances, they inevitably led to the fulfillment of the challenge faced by all the rocket societies of the 1930s, the harvesting of rocket power for the exploration of the heavens.[128]

VI

The American Rocket Society

The Science Fiction Writers Period—The American Interplanetary Society

Science fiction was the real parent of the American Rocket Society, because on the evening of 4 April 1930 the 11 men and one woman who gathered in the third floor apartment of a small brownstone building at 450 West 22nd Street in New York City to form the American Interplanetary Society (later the American Rocket Society) were almost all members of the fraternity of writers who contributed to Hugo Gernsback's *Science Wonder Stories.*

The apartment was the home of two of the people present: G. Edward Pendray, a *New York Herald Tribune* reporter who wrote for Gernsback under the pen name of Gawain Edwards, and his wife, then a widely syndicated woman's page columnist for United Features who had her own *nom de plumes.* Mrs. Pendray was well-known under her maiden name of Leatrice, or Lee Gregory. Also present at the meeting was David Lasser, the moving force behind the movement and the man chosen to be the Society's first president. Charles P. Mason, Charles W. Van Devander, Fletcher Pratt, Nathan Schachner, Laurence E. Manning, William Lemkin, Warren Fitzgerald, Adolph L. Fierst, and Everett Long (or Roy A. Giles—the record is not clear), completed the group. Of this dozen, nine were either science fiction writers or associated with the Gernsback organization. The remaining three were undoubtedly science fiction buffs.

Lasser was the managing editor of *Science Wonder Stories* and had obtained a Bachelor of Science and Administration degree at the Massachusetts Institute of Technology. He had worked only briefly in engineering and industry. Charles P. Mason was the Associate Editor of *Wonder Stories* who submitted his own stories under the delightful pseudonym of "Epaminondas Snooks D.T.G." Van Devander was another journalist who also wrote science fiction on the side, under the name "Peter Arnold." Fletcher Pratt was a prolific writer both of science fiction and history, including a series of books on the Civil War, a series on the Napoleonic period, a biography of James Madison, and also works on naval affairs. Nathan Schachner, a lawyer by profession, writing under the name of "Nat Schachner," was one of Gernsback's most popular authors. Manning was starting as another Gernsback writer. His first story, written with Pratt, was to appear in *Science Wonder* for May 1930. Dr. William Lemkin, the only Ph.D. of the founders, was a Russian-born chemistry teacher and textbook writer employed at New York's High School of Commerce. He too wrote science fiction on the side, contributing half a dozen stories to Gernsback's magazines. Fierst wrote no stories himself but was Gernsback's re-write man.[129]

Warren Fitzgerald is said by science fiction historian Sam Moskowitz to have been the first and last black man to have actively engaged in science fiction "fandom," or fan clubs of the 1930s and was the first president of the New York club, The Scienceers. Mort Weisinger, one of the original members of The Scienceers, adds that this group was originally to have been called The Rocketeers, though no experiments were planned.[130]

There is some doubt about the identity of the twelfth founder of the Society as Everett Long is usually credited, yet his name does not appear on the earliest membership list. Roy Giles' name does and is included with the eleven other founders. Nothing is known of Long, but Giles was a reporter and free-lance writer, professions which certainly fit the pattern of the typical AIS founder.[131]

Hugo Gernsback himself joined the Society, but is not known to have attended meetings. More than likely he was seeking interplanetary story ideas through Society literature. The Gernsback entourage, however, was a magnet that attracted still other science fiction fans who were space-minded. An example was Dr. Samuel Lichenstein, a dentist who wrote at least one story for Gernsback. Lichenstein, who was Schachner's brother-in-law, became the Society's long-standing and reliable treasurer after he assumed the post from the first designated treasurer, Laurence Manning. Pendray afterwards acknowledged that the Society owed a great deal of gratitude to Lichenstein who very faithfully and competently undertook his duties.[132]

Other posts filled at the opening meeting 4 April 1930, were the Vice-Presidency which went to Pendray and the Secretaryship that went to Mason. The historian, Fletcher Pratt, was named the Society's first librarian. The newspaperman and future press secretary of New York Governor Averell Harriman, Van Devander, was selected editor of the Society's planned organ, the mimeographed *Bulletin of the American Interplanetary Society.*[133]

The same infectious optimism and naivety that characterized the German Rocket Society in its earliest days also pervaded the Americans. "We believed generally that a few public meetings and some newspaper declarations were all that would be necessary to bring forth adequate public support for the space-flight program," remembers Pendray. "As for technological help, it was our expectation that engineers and scientists would spring to our service if we but called their attention to the possibilities of rockets in an appropriate manner." The aims of the new Society, as stated in its constitution, likewise echo the high ideals found in the VfR's founding charter.[134]

The protocols and by-laws of the Society's constitution were comparable to the VfR's (there are stipulations for a Board of Director of five members in good-standing, and so on). Revenue was obtained through annual dues: ten dollars for active, and three dollars for associate members. An amendment of 1 October 1932 made an allowance for members living outside New York, providing for an expansion of the Society. Dues were $7.50 per year for those residing within a radius of 100 miles of the city, and $5.00 a year for those living beyond.

Membership numbers of the American Interplanetary Society (or the ARS) did not match those of the VfR. Pendray guesses there were about 15—20 members in the first few months of operation. Great hopes were focused on Alfred H. Best when he joined about 1932, as it was known he was a scion of the wealthy family that ran The Best Pencil Company of New Jersey. Other than the customary dues, however, no money was forthcoming from that quarter. By 1935 the claimed membership was 300. "But this was always said with tongue in cheek," says Pendray, as "most of these 'members' didn't pay their dues." In 1940 the number perhaps reached 400. As with the VfR, an occupational breakdown would be most intriguing. Spread out over a period of years it is readily apparent that writers made up the first wave, while the professional engineers came in a trickle thereafter until their numbers grew far greater than those of writers in the immediate prewar years. The Depression too had its impact. Generally, the unemployment rate for degreed engineers was far less in this country than it was in Germany, which had a smaller industrial base. Hence, the VfR found no shortage of qualified and unemployed technicians with a lot of time on their hands. In the United States the situation was reversed. There were relatively few unemployed engineers who could—or would—devote full working days to rocketry as in Germany. This is one reason why the emphasis during the first two years of the American Interplanetary Society was less technological than propagandistic.

In a vague sort of crusade, the Society first tried to "educate" itself through its so-called library that meant highly popularized articles rather than books. Fletcher Pratt's "Librarian's Report, 1930—31," flatly states in his introductory statement that "in the course of its first year the library of the Society has made little progress."[135]

To attract attention to the Society's first public meeting on 30 April 1930 at the American Museum of Natural History, the noted polar explorer Captain Sir Hubert Wilkins "donated" a leather-bound copy of *The Discovery of a New World, or a Discourse Tending to Prove That There May Be Another Habitable World In The Moon* (1640), written by Sir Hubert's illustrious ancestor, Bishop John Wilkins of Chester. Resulting publicity, however, caused hardly a ripple, even though Wilkins' name was then prominent in the papers since he had recently returned from a news-making voyage to Antarctica. The book had actually been purchased by Lasser.[136]

In June 1930 the *Bulletin of The American Interplanetary Society* first appeared. It announced the Wilkins donation and also that Charles P. Mason had been entrusted with making a survey of the "entire field of information relating to interplanetary travel." Fletcher Pratt's paper on "The Universal Background of Interplanetary Travel," which had been read before the Society on 2 May, was also printed as well as the prediction by Princeton University's Dr. John Q. Stewart that a 33.5 meter (110 foot) diameter metal sphere propelled by ionized hydrogen, manned by a crew of 60 men and a dozen scientists, would reach the moon by the year 2050. Another prediction was offered, made by Esnault-Pelterie who envisioned a trip to our satellite not in 120 years, but 15. The delay, he cautioned, would be caused by the *sine qua non* of undertaking costly experiments, perhaps $2,000,000 worth. European experiments—and disasters—were also noted. Valier's death on 17 May was reported, as well as the mail rockets of the Czech inventor, Ludvík Ocenásek.[137]

America's Robert H. Goddard was not forgotten. His name became the most often cited in *The Bulletin* and its successors, *The Journal of the American Rocket Society,* and *Astronautics.* Just four days after the famous 4 April founding meeting at the Pendrays', Lasser wrote to Goddard soliciting his support and asking him to give a talk. He politely declined.[138]

Goddard's refusal to become directly involved with Society affairs set the tone of his relationship to the Society for years thereafter. But he did become a member. From Worcester he responded to Lasser on 12 April: "My position is a peculiar one. So far as I know, I am the only one who has worked out the problem in sufficient detail to know the general principles which must be employed in its solution, and who has at the same time checked conclusions by actual experiments, certainly in this country and as far as I have learned, abroad also. I do not feel in a position to present the whole story yet, and any talk, however informal, is bound to lead to questions which will leave matters in an unsatisfactory state if they are not answered. I have, for this reason, declined several attractive opportunities to lecture and to write on the general subject of rocket propulsion in the last few months.''[139]

In this self-imposed ivory-tower seclusion from Society affairs, Goddard was not alone. He had much in common with his contemporaries in Europe—Oberth in Germany and Tsiolkovsky in Russia. Of the three, Oberth was willing to cross the line and become more public. He even shared in the conduct of the German Society but was self-admittedly an inept manager and soon shied away when things did not go right. In short, none of the three were organizational men in either disposition or inclination. Lasser was not to contact Goddard again until almost a year later.[140]

In the meantime, the Society could do little more with their paltry funds and handful of members than hold bimonthly meetings in the American Museum of Natural History and to publish what they could in the *Bulletin* and popular scientific press. All attention was thus riveted upon European activites. The REP-Hirsch Astronautical Prize was noted, as was Oberth's planned 21 kilometer (13-mile) sounding rocket with red tail light for telescopic tracking. Most important was Willy Ley's letter to the Americans informing them of the establishment of the *Raketenflugplatz* in September, and the first experiments of Riedel and Nebel. The Europeans were early aware of the existence of the American group but probably considered them ill-grounded in theory and, with the exception of Goddard, years behind. However, Werner Brügel invited Secretary Mason to write a brief description of the goup for his *Männer der Rakete,* printed in 1933. Some Americans and Europeans hoped for a universal (i.e., one-world) space expedition representing the planet Earth, but nationalistic pride was still in evidence. ''Details about these experiments,'' wrote Ley to Lasser,'' will be given in our next [VfR] bulletin, but we merely want to inform America by these few lines that the German Interplanetary Society [sic] has not gone to sleep and that our work has brought the world another step nearer the final goal. This goal is the space ship, as our president, Prof. H. Oberth has shown in the film for which he was advisor, on 'The Girl in the Moon.'.''[141]

The Americans would soon attempt to capitalize upon this same movie. They also attempted to host one of Europe's deans of astronautics, Robert Esnault-Pelterie, who was commonly called REP by his acquaintances. That a space travel group now flourished in America excited all of the pioneers dedicated to the great dream of space travel. Every convert was progress, and the more widespread the idea, the greater the chances were of raising the necessary astronomical capital. Esnault-Pelterie therefore not only kept in touch with Goddard but with David Lasser as well. In June he had submitted both an application for active membership in the Society and also an autographed copy of *L'Astronautique,* which had just come off the press. REP went one step further and ceded the American rights of his book to Lasser's organization. An English translation was contemplated but never made. REP had one more surprise. He expected to visit the United States in the fall to pursue a patent suit on his airplane ''Joy Stick'' invention. Upon confirmation of this visit, Lasser cabled the Frenchman inviting him to address the Society at the Museum. To his astonishment, Esnault-Pelterie accepted. Lasser immediately became busy arranging for the hall and generating the publicity. The January 1931 issue of the *Bulletin* announced that REP would speak on 28 January ''at the first large public meeting sponsored by the Society'' and that he would be met by officers of the Society upon his arrival on 15 January. The hall held about 1,000 seats and admission was to be free. On the 12th Lasser had a brainstorm. He telegrammed Goddard, inviting him to attend, but Goddard again refused him.[142]

With or without Goddard, Lasser would make a coup far greater than the now all but forgotten Wilkins ''donation'' 11 months earlier. Lasser was not sure that even with this billing the auditorium would be filled. Space travel was still an arcane subject and the Society, in Pendray's words, was ''unknown and impecunious.'' Lasser had another brainstorm. He telephoned Ufa's New York office and obtained an English-language copy of the film *Frau im Mond.* The Executive Commitee excised the drawn-out romantic scene and left only the technical portions, which were dramatic and artistic enough. Publicity was posted everywhere. The Musuem authorities now became anxious that the event would draw too many people. Extra guards were hired, says Pendray, thus ''inflicting almost mortal injury upon the Society's treasury.'' Besides the pecuniary damage, another disaster struck. Esnault-Pelterie had arrived on time and worked with the Society's officers on his speech but suddenly could not appear. He had developed a severe cold that confined him to his hotel room and he notified the group just four hours before the scheduled time of the meeting. At a quick conference it was decided that one of the members of the Society would have to take his place. "I was the man elected to do it," recalled Pendray 25 years later. "Lasser, as chairman of the meeting, announced the substitution in an unmistakable way, but nevertheless many members of the audience came clamoring up to the rostrum afterward demanding Esnault-Pelterie's autograph. After several fruitless efforts to explain that I wasn't the great Frenchman in person, I gave up and signed Esnault-Pelterie's name right and left. As a result of that night's work there are hundreds of copies of Esnault-Pelterie's signature in autograph collections today that couldn't be phonier." Pendray has opined that his Van Dyke beard, which gave him a very continental look, may have had much to do with his monumental mistaken identity.[143]

The February 1931 issue of the AIS *Bulletin* announced on its front page the huge success of the meeting, that an estimated 2,000 persons had shown up, that Dr. H. H. Sheldon of the Physics Department of New York University had presented an outline of Esnault-Pelterie's ideas, and that the Frenchman unfortunately could not attend. Nothing was said of Pendray's unintentional doubling for the Frenchman. The size of the crowd, it was also reported, lived up to the apprehensions of the Museum staff. The entire program had to be repeated for those who stood in line outside for almost two hours. Esnault-Pelterie's speech was also printed in that issue of the *Bulletin*.[144]

The Society now numbered 100 and had been fully launched. The 1 March 1932 edition of the *Brooklyn Daily Eagle* announced a way that this small number could spread its voice and image to a considerably further distance—TV. G. Edward Pendray was to broadcast his talk "The Rockets of the Future," simultaneously over W2XAB, Columbia Broadcasting System's experimental television station and W2XE, Columbia's "World-Heard Short-Wave Station." This was the 37th in a series of "aviation talks" sponsored by the *Eagle*. It surely must have been the world's first TV space report.[145]

A more accurate gauge of the Society's impact upon the public is found in the "Report of the Secretary—1930—1931," signed by C. P. Mason. Mason meticulously compiled a month-to-month tabulation of "Inquiries" and the number of new members gained from the inquiries. (The new members he designated as "Associates" to distinguish them from "Active" members, or those people from the immediate New York area who could show up at meetings and later involve themselves in experimentation.) From April 1930 to March 1931 Mason counted a total of 103 inquiries received and answered, with a gain of 39 associates and one transference to active status. On a geographical breakdown, the results were even more interesting but also as expected. New York and suburban New Jersey took the lead with 26 inquiries and 18 associates, total. The remainder of the states, including the District of Columbia and Hawaii, averaged two inquiries and one associate. Mason also provided a column for "At Sea" which amounted to one inquiry and one associate. This must have been future science fiction writer Robert A. Heinlein, then an ensign, who mailed in his application aboard the USS *Lexington* in the Pacific. Of the foreign countries only four are listed: Canada, Mexico, Russia, and Honduras.

American scientists were also increasingly taking space travel seriously. From the earliest days, however, the scientific community—primarily physicists—was well represented among the subscribers to the *Bulletin*. A "Supplementary Mailing List for [the American Interplanetary Society] Bulletin" found in the Pendray papers lists some ten professors and four individuals titled as Ph.D.'s. The two most notable ones are Dr. Alexander Klemin of the Guggenheim School of Aeronautics and afterwards to become one of the most outspoken and active of the Society's learned members; and Professor Percy M. Roope, Goddard's associate at Clark University, Worcester, Massachusetts. Goddard was considered a regular subscriber. Also on the Supplementary Mailing List were a number of public libraries.

Though these entries seem impressive, there were few from the ranks of American academia who openly espoused the cause of space travel in 1930—1931. The fantasizers and publicists made up the core of the movement. Lest they risk their professional reputations by being associated with a crank cause, the scientists generally felt it wiser to remain adherents *sub rosa*. Still, there were at least three who dared come into the open. Lasser's campaign appeared fully vindicated. Dr. Harold Horton Sheldon was probably the third Ph.D. member after William Lemkin and Dr. Clyde Fisher of the American Museum of Natural History, through whom the

Society had obtained their meeting place. As science editor for *The New York Tribune,* Sheldon was a powerful voice in the promotion of space flight at the time. He also gave his own lectures and wrote the foreword for Lasser's book. Sheldon's specialties were many, including the conduction of electricity, the theory of crystals and photoelectric color measurement. Even if these abstruse sciences were not immediately adaptable to the space ship, Dr. Sheldon lent a greater service to the Society and to the movement, that of respectibility. Another learned member, Dr. Alexander M. Zenzes, who lived in San Francisco at the start of 1931 was attempting to bring out the works of Max Valier and Hermann Oberth in English. These projects, however, never materialized. The charter member Dr. Lemkin, who was a chemist, contributed articles on rocket fuels. Lasser was also publishing in *The Scientific American.* Lasser felt so confident in the apparent progress that in the first annual President's report he proposed a Committee on Experiments and the creation of an International Interplanetary Commission. The utopian aspect of the space flight dream had become almost an obsession. "There should be," he said, "an international press service where news on world developments will be gathered and translated into the various languages and disseminated to the various national societies. The International Commission would also cooperate in the actual fulfillment of our goals to the fullest extent possible. I can foresee the building of the first space ship only as a joint effort of an united earth." It is ironic that extreme pessimism rather than this altruistic optimism was the real motivating factor for at least one, and perhaps other early members. Bernard Smith, who joined about 1932 after reading of a coming meeting at the Museum, recalled, without tongue-in-cheek that the Depression years for him were anything but uplifting. He was already poor when the stock market crashed and as an itinerant handyman was constantly out of work. "It was a lousy planet," he felt, and "the rocket ship was the only way to get off it."[146]

Smith enlisted in the Society at the right time. His mechanical aptitude was sorely needed for building the first rockets. The inspiration to make them had always existed but few technical people were available, only dreamy-eyed writers. The technical inspirations came both spiritually and concretely from Germany.

Early in 1931 the Pendrays planned to go abroad. The trip was planned in such a way as to enable them to pay visits to the European experimenters. Mr. and Mrs. Pendray were designated the Society's official representatives. "But in view of the state of the treasury," wrote Pendray, "we paid for the trip ourselves." Pendray had been re-elected Vice-President of the Society (Lasser remained President) that April, and in the same month they departed for Europe on a brief Mediterranean cruise aboard the *Ile de France.* The first stop was Naples. Here Pendray had hoped in some way to learn more about a certain Dr. Darwin O. Lyon who had been making some fantastic claims in the papers. In a December 1930 dispatch from Vienna, Lyon had asserted that he had completed his beryllium alloy two-stage liquid fuel rocket of 59.8 kgs (132 pounds) which he was about to shoot to a height of 113 to 145 kms (70 to 90 miles) from atop Mount Redorta in the Italian Alps. In February came the word that the rocket had exploded, injuring the experimenter and several mechanics. The ensuing adventure of the shadowy "physicist" continued in the pages of the *Bulletin* as well as in *The New York Times* and in papers around the world. In July, after Pendray had returned to America, Lyon further announced that he had moved his launch site to Tripoli. Instead of instruments, he would now send up "two birds and two mice, for the purpose of studying their reactions under the influence of cosmic rays." In September, Lyon reported that he had already sent the animals to more than 1.6 km (one mile), and that a more ambitious rocket was being prepared in Paris. Pendray had previously written to Lyon at Milan, the closest town to Mount Redorta. None of the letters were answered. Pendray had his suspicions. Lyon, it would appear, was nothing but a fraud attempting to extort money from several countries for his supposed research. Not one rocket had been built. All the data were extrapolated from the open literature. Pendray was not to see Lyon in Italy. From Naples the *Ile de France* sailed to Marseilles and from there the Pendrays took a train to Paris. Phone calls to locate Esnault-Pelterie proved fruitless. The great French pioneer was undertaking his first experiments, one of which, on October 9, cost him four fingers of his left hand when some liquid tetranitromethane exploded. He was, at any rate "very busy." From Paris the Pendrays went to Berlin. Here their effort more than paid off. Pendray had no difficulty meeting his counterpart, Willy Ley, the Vice-President of the *Verein für Raumschiffahrt,* whom he facetiously called "the G. Edward Pendray of Germany."[147]

They had shared correspondence ever since the formation of the American Interplanetary

Society, although Ley knew little English and Pendray's German was equally non-fluent. Technical conversations were almost impossible, save through technical drawings. However, Ley was a gracious host and took the Pendrays to the VfR's new testing field, the *Raketenflugplatz*. The firing of a small Repulsor motor was the highlight of their trip. It was the first liquid-fuel motor firing they had ever witnessed and the first ever built to their knowledge, as they were unaware of Goddard's successful launches since 1926. "It filled us with excitement," Pendray recalled, "and upon our return we reported fully to the Society, on the evening of 1 May 1931, both the method and promise of the German experiments."[148]

That same May, the Society incorporated under the laws of the State of New York. It was also in May that Pendray published a full report of his findings in Germany in the *Bulletin,* mistakenly asserting that no successful liquid-fuel rocket had ever been flown. To this, Goddard reacted swiftly and once and for all established his priority. His response was published *in toto* in the June-July 1931 issue of the *Bulletin.* Goddard was more than ever convinced his aloofness was justified. To Pendray and his fellow members, the incident seemed to confirm the eccentricity of their idol—Goddard—and the unreasonableness of his stand. Pendray himself again wrote to Goddard at this time to urge him to reconsider participating in the Society, but Goddard remained unmoved, and he reiterated his position.

In later years Pendray reflected: "We couldn't get Dr. Goddard to throw in his lot and come along but I guess we had more to gain from him than he had from us. I still wonder, though, if he might have gone even further if he hadn't insisted on going alone."[149]

The Experimental Period—The American Rocket Society

Pendray and the rest of the American Interplanetary Society were particularly anxious to obtain Goddard's cooperation in the summer of 1931 because they were preparing to set out upon their own experiments. The major problems were common to all experimenters: money, a suitable place to fire the rockets (an especially acute problem in New York), and technical expertise. The September issue of the *Bulletin* outlined and illustrated a preliminary rocket test stand. The tests would all be based upon liquid oxygen and gasoline. Following the German example, the motors would also be water-cooled. An open solicitation was made to interested, and, it was hoped, talented members to submit further information or their time to begin the work. As it turned out, the cost was far more reasonable than Pendray's Experimental Committee had estimated. Instead of the contemplated $1,000 they thought they needed, the actual cost amounted to a grand total of $49.40. This was possible because of "donations." Fifteen litres of liquid oxygen was furnished free by Dr. George V. Slottman, an officer of the Air Reduction Company. The Air Reduction Company's repeated generosity prompted Pendray to remark to the Treasurer that, "the friendliness of this company toward our Society and its endeavors deserves comment. I think we should organize a subsidiary of the American Interplanetary Society, to be known as the Air Reduction Extolling Corporation, the members of which are to be pledged to extol the Air Reduction Corporation [sic] at all hours of the day and night, when not otherwise engaged." The castings for the motor were also provided without charge, by John O. Chesley, manager of new developments of the Aluminum Company of America. Chesley had three rough duralumin blanks made from a pattern furnished by the Society which were then machined and drilled in the shop of one of the new AIS members, Hugh Franklin Pierce, later President of the Society. Valves also cost nothing. Smith recalled that he obtained them by posing as an "interested manufacturer" and asking for a "sample." It is no wonder that Pendray jotted down 50 cents for "Valves, etc." in his careful tabulation of the component costs of what became known as "American Rocket Society Rocket No. 1," or ARS Rocket No. 1. The aluminum water-jacket surrounding the motor was a cocktail shaker given away, as a premium by a chocolate milk company. The parachute was a five-dollar piece of silk pongee purchased from a department store and cut and sewn by Mrs. Pendray after she and Pierce had studied a large man-carrying parachute at an airplane exhibit. The parachute holder was a small aluminum saucepan minus its handle. A spring was added for ejecting the chute. Through Slottman's largesse, the Society also purchased a second-hand oxygen container from his liquid air company for $15 (they cost about $100 new), and he also shipped a tank of compressed nitrogen across New Jersey to the proving ground, charging the Society only the transportation cost. As for locating the launch site, Pendray and others roved about the countryside for many months during which they inspected army fields, grasslands, waterfronts, and even dumping grounds for refuse. Chemist Dr. Lemkin was the "chauffeur."

Eventually, in the summer of 1932 an old farm near Stockton, New Jersey, facing the Pennsylvania border near the Delaware River, about 60 miles out of New York was deemed suitable and found to be available. Ace Hewitt, the owner, granted permission. A technical crew was not so easy to find, but luck was with them in this department as well. In Hugh Franklin Pierce, the Society found excellent help. Pierce's Depression job was selling tickets on the New York City subway. He purchased a few tools which he set up in the basement workshop of his Bronx apartment at 501 E. 136th Street. Ohio-born Pierce had little formal schooling, but had natural mechanical ability and many months of training as a machinist for the Navy, a course that he never completed. On 27 October 1931, Pierce wrote to Pendray to inform him that: "I have obtained the use of a basement work shop which is ideal for our purpose. There will be no rent for one year, the only expense will be electric light and power. I have found a lathe that is in A-1 condition for $55. This is a portable machine. I have brought a set of dies & taps for thread cutting for use in this work. These tools I wish to keep as my own property. If the Society does not wish to purchase this lathe, I will buy it as my own to be used in these experiments. If this is satisfactory please let me know." Apparently it was. Based upon Pendray's German report and suggestions from other members, Pierce built the first rocket himself.

In the December number of the *Bulletin,* Pendray was able to report that: "I am able to announce on behalf of the Society, that we are now actually building a small rocket—a preliminary experiment which we expect will lead soon to much more important ones. The rocket will probably be completed in about a month, if all goes as planned. It will be a rocket of the two-stick Repulsor type, standing about six feet high [1.8 meters], and will be equipped with an automatic parachute, though it will probably not develop sufficient lift to carry any instruments . . . We will use as fuels liquid oxygen and high-test [ethyl] gasoline."[150]

When completed, the rocket consisted of two parallel aluminum tanks 1.7 meters long (5.5 feet) and 5 cms (2-inches) in diameter. One contained the fuel, the other the oxidizer. The gasoline was forced into the combustion chamber by nitrogen gas under 21 kgs/sq/cm (300 psi) pressure, the oxygen by its own pressure produced by partial evaporation. Overall diameter of the rocket was 20 cms (8-inches), and 2 meters (7 ft) in height. The engine itself was an aluminum casting of 7.6 cms (3-inches) diameter. Loaded with propellants, the rocket weighed but 6.8 kgs (15 pounds), complete with two sheet aluminum stabilizing fins. Almost as soon as it was ready, the rocket was carried to the Washington Square Campus of New York University where it was demonstrated in a lecture by Pendray, Pierce, and apparently Dr. Sheldon. It was also shown at the Society meeting at the Museum of Natural History on the evening of February 18. Pendray and Pierce demonstrated that parachute release mechanism operated by nitrogen pressure and the valve systems, by forcing in air from a tire pump instead of the oxygen and gasoline. "The small rocket we show you tonight," Pendray concluded, "is a start in the direction of interplanetary flight. . . . A thousand members will permit us to continue these experiments. . . ."[151]

The attempted shot itself did not occur as expected. Pendray admitted the rocket had not been designed for high altitudes but merely to test the value of the design. He still entertained the hope that "a height of several miles may be achieved" and that "If this occurs, the rocket will have the distinction of the altitude record for all liquid-propelled rockets. . . . It is hoped that the parachute device in the rocket's nose will function satisfactorily so that the rocket can be retrieved and shot a number of times, and its full value can be estimated." Not until 12 November 1932 was it taken out to Stockton for ground tests prior to the anticipated launch. The delays were caused by "laboratory tests" and searching for the site. The July 1932 issue of *Astronautics,* the successor to the *Bulletin,* thus reported that, "After considerable search for a field in the environs of New York in which to shoot the experimental rocket constructed by the Society, a place has tentatively been found in the neighborhood of Stockton, NJ. . . . If the tests are successful the rocket will be shot about October 10." The laboratory tests consisted of pressure-testing everything. Beginning late in August, a small group of members of the Society, including charter members Lasser, Schachner, Lemkin, Manning, Pendray, and Mrs. Pendray, along with newer members Pierce, Alfred Best, and Alfred Africano, prepared two bomb-proof dugouts with sandbags. Other members built the wooden launching stand. Pierce put together an electrical ignition system. Following all the suspense of valve and ignition trials, the rocketeers returned to Stockton Saturday morning, 12 November, in cold and rainy weather for a possible launch the following day. The experiment on the 12th was a static trial for combustion and thrust qualifications. "At about five in the afternoon," Pendray reported soon afterwards, "we are now ready. Mr. Pierce threw his switches rapidly. The fuse apparatus worked to perfection. For an instant there was a great flare as the pure oxygen struck the burning fuse [the electrical system had failed and was discarded]. In an instant the gasoline was also pouring into the rocket. The fuse, the flare, and the uncertainty about the performance of our rocket motor all disappeared at once, as, with a furious hissing roar, a bluish white sword of flame shot from the nozzle of the combustion chamber, and the rocket lunged upward against the restraining springs."

For 20 to 30 seconds a thrust of 27 kgs (60 pounds) was reached. "Suddenly," continued Pendray, "we knew that the oxygen supply had been exhausted. There was an excess of gasoline, as we had planned. This now came spurting out, throwing a shower of fire all around the foot of the rocket and proving stand." After the fires had died down and the damage assessed, the verdict was that the rocket was not fit for a flight and the difficulties met in getting all the parts to operate satisfactorily at the right time pointed to such a total reconstruction as to constitute making it into a new rocket.[152]

There had been changes within the Society itself that year. More or less permanent office space (at least until 1936) was available from the Milk Research Council, not out of the dairy industry's concern for the exploration of extraterrestrial spheres but because Pendray was employed as the Milk Council's editorial editor from 1932 to 1936 and had agreed to let him conduct Rocket Society business in his office. The Society had no other office at the time. The Milk Research Council, or "Milk Fund," as it was sometimes called, otherwise had nothing whatever to do with rockets nor space flight. It was an organization supported by the major milk companies of the New York-New Jersey area during the Depression to promote the use of milk in low-cost diets. As for other changes in the Society, the founder and first president, David Lasser, had relinquished his chair to Pendray in the third annual meeting of 2 April 1932. The lawyer, Nathan Schachner, was voted Vice-President.

To the library could now be added Lasser's own book, *The Conquest of Space,* released by Penguin Press in late September 1931, making it the first English-language work on the subject. In a sense the Society was responsible for the book. Penguin was a private or "vanity" press, not to be confused with the present publisher of that name, established for the sole purpose of publishing Lasser's work and was financed by AIS members Lasser, Pendray, and Schachner. All three were aware that there simply was no book available and they also wished to further propagandize or "educate" the public in space flight. Lasser was the logical choice for authorship as he was both President of the Society and a professional writer with *Wonder Stories.* The difficulty was that no publisher was interested. The three men consequently pooled about $12,000, forming Penguin Press. About 5,000 copies were printed, Lasser recalls, and there were perhaps as many in the British edition, published in 1932 by Hurst & Blackett of London. A fourth AIS member, Dr. H. H. Sheldon of the American Museum of Natural History added greater prestige to the book by writing a forward.

It was about this period, however, that Lasser found a greater personal challenge than space-flight. He wished to do what he could to alleviate the economic ills of the country and shortly resigned his Presidency of the Society to embark upon a career as a trade union organizer. Pendray became the second President in May 1932. That same month the old mimeograph *Bulletin of the American Interplanetary Society* became the new, offset printed *Astronautics,* with Lasser remaining editor until at least April 1933. By 1932, according to the journal, AIS membership was divided over 21 states and nine foreign countries. The Russian members were Rynin and Perelman.

On 18 June the Committee on Biological Research, which consisted of Laurence Manning, the nurseryman and popular science fiction author, with one Thomas W. Norton, reported their findings of subjecting two hapless guinea pigs in a .6 meter (two-foot) centrifugal drum revolving at 600 rpm to a force of about 30g's. Both test subjects died. They were thoroughly autopsied, the principle cause of death being clotted blood in the hearts which could not cope with the increased weight of the blood. ". . . In the case of a human," they concluded, "our experiment tends to indicate that man will not be physically capable of withstanding very much higher rates of acceleration than those limits now set by present physiological theory."

Unknown to Manning and Norton, the students Wernher von Braun and Constantine Generales had conducted similar centrifugal experiments about the same time upon mice, to 220 g's, in the corner room of a house near the University of Zürich. (Norton had actually predated the German and the Greek by a year when, as an American Interplanetary Society project, he worked with a Konrad Schmidt and subjected white mice to 80 g's.) Earlier, unknown to them all, the Russian Rynin and several physicians had actually performed an even wider range of centrifugal trials in the spring and summer 1930 upon a menagerie of insects, fish, birds, pigeons, mice and rats, cats, rabbits, and even raw eggs! If anything, these exotic space-oriented physiological investigations show that there was a woeful scarcity of open literature at the time or a genuine scientific exchange world-wide.[153]

G. Edward Pendray became so engrossed with experimental rocket No. 2 that he resigned as President half way through his term, ceding the chair to the Vice-President, Nathan Schachner. Manning became the new Vice-President. Pendray wished, he said, "to devote his time exclusively to rocket experimentation rather than to administrative duties." At the time the technical problems seemed formidable but now would be considered elementary. Bernard Smith became the principle architect of the new vehicle. One day the young, unemployed jack-of-all-trades, later to become Technical Director at the Naval Weapons Laboratory at Dahlgren, Virginia, showed up at a post mortem meeting on the first rocket and began criticizing its flaws. Pendray's eyes went up. Here was someone who obviously knew what should be done. "How about making a new engine?" Pendray asked. "Sure," came back the response. The engine was built and flown. Smith salvaged the tanks of ARS Rocket No. 1, disposed of the water jacket and the parachute and clamped the motor securely in the upper portion between the two propellant tanks. The rocket was thus still a nose-drive affair, an unsuitable arrangement that did not guarantee stability as assumed. Nonetheless, it was carried over from the Germans who thought it the only workable means.

Smith also substituted light balsa-wood fins for the aluminum vanes and generally streamlined the rocket. A hole was cut into the nose for air cooling. ARS No. 2, as it was called (technically, it should have been designated AIS No. 2), was shot off on the beach at Marine Park, Great Kills, Staten Island, on Sunday morning, 14 May 1933. It rose abruptly to approximately 76 meters (250 feet) for two seconds, then came crashing down when the oxygen tank burst with a loud popping sound. No matter, it was still considered a high success; it was the first flight of an American Society rocket.[154]

As there had been no parachute (omitted to save weight), the launcher was tilted five degrees to seaward where the rocket was expected to land for safe recovery. In actuality, the rocket plummeted into the water of lower New York Bay where it was fortuitously rescued by two boys who happened to be in a rowboat nearby. The launching itself had been swift and dramatic. After rehearsals and a sort of countdown, the "Valveman" Smith attempted to pull the cord which opened the rocket's valves prior to ignition. A detachable lever, probably loosened by the stiff wind which had been blowing that day, fell off. Smith courageously sprang forward with a hastily improvised torch and ignited the fuses, not having enough time to return to his shelter before the rocket took off. On hand to witness the feat were not only the jubilant launch crew, consisting of Pendray ("Command"), Smith, Manning, Alfred H. Best, Carl Ahrens and Alfred Africano, but also the new Secretary, Max Kraus; Schachner; Daniel De V. Harned, a nonmember who had local political connections and who was responsible for obtaining permission to use the ground; representatives of the Bureau of Combustables of the New York Fire Department; cameramen from Acme News Pictures and Universal Newsreel; and assorted spectators including the two boys in the boat. The next step was, of course, a new rocket.[155]

The fourth annual meeting saw Laurence Manning elected as the new President, with Mason as Vice-President, and Dr. Lichenstein with two jobs, Secretary and Treasurer. Lichenstein could sadly report a near insolvent financial posture of the Society, but somehow they overcame it. The American Interplanetary Society's total liquid assets were $15.37. With this money and sheer bold determination, President Manning announced that not one but three new rockets were planned, though members were temporarily to be deprived of Astronautics to recoup some of the money. The three designs were promptly approved, with rocket No. 4, constructed by professional civil engineer John Shesta, being completed first. It became the next to fly but only after an initial failure.

Shesta, a Russian-born member who obtained his civil engineering degree from Columbia

University in 1928, had designed a novel motor featuring four nozzles protruding from a single chamber. The rocket still retained the favored nose-drive arrangement but was otherwise far sleeker in appearance than the previous rockets as the tanks were placed in tandem below. The nozzles were so arranged as to direct the jets outward slightly from the vertical. Shesta returned to a water jacket for cooling, and included a parachute. The first attempt to launch this rocket was likewise made at Staten Island. On that particular 10 June Sunday morning, the fuel inlets proved to have been made too small and the rocket simply did not build up enough power to rise. After appropriate modifications, a second attempt was made 9 September 1934. The flight was considered by the Society's Experimental Committee "one of the most successful and spectacular shots ever obtained with a liquid fuel rocket." Careful calculations on a special triangulation system showed that ARS No. 4 soared to an altitude of 116 meters (382 feet) and landed 407.8 meters (1,338 feet) from the base of the new adjustable metal launch rack and covered a distance of 483 meters (1,585 feet). The rocket's greatest velocity was calculated at more than 304.8 meters/sec (1,000 feet/sec). The parachute failed to open because the horizontal flight prevented operation of the release mechanism, but otherwise it was a success.[156]

Rocket No. 3, designed by Smith and Pendray, was also innovative. The tanks were nestled one inside the other, with the gasoline tank inside and the oxygen on the outside. The motor was mounted on top as usual but had a very elongated nozzle that ran through the inside length of the gasoline tank. The primary objectives of this construction were to keep the oxygen tank away from the rocket's flame, to test the cooling potential of gasoline surrounding the motor, and to determine the efficiency of long nozzles. It became impossible to fuel or launch this bulky missile because the liquid oxygen in the outer tank was constantly exposed to so much warm metal it evaporated and blew out of the top fillhold as fast as it was poured in. No. 3 consequently, was never flown. No. 4, in fact, was the second and last of the Society's liquid-fuel rockets to fly. There were solid-propellant flight shots later, but the rest of the experimental work was conducted essentially with an eye towards scientifically perfecting the motors on test stands.[157]

The Society's embarrassing financial situation necessitated the abandonment of further flight tests. For the same reason there was a lapse of some seven months in the publication of *Astronautics.* An important block of history of the Society is thus lost, save for the recollections and private papers of Pendray. Virtually unknown, for example, is Pendray's ambitious "Proposal for the Establishment of a Fund for Rocket Research with the Object of Developing High-Altitude Rockets for Scientific and Meteorological Investigation," to be sponsored by the American Rocket Society, the Guggenheim School for Aeronautics, the Smithsonian Institution, and the United States Weather Bureau. Pendray hoped to solicit $100,000 from those institutions for the development of, initially, weather rockets. The work was contemplated to take five years with monies allotted for research, salaries, and facilities. In February 1934 the plan was submitted to Alexander Klemin, a good friend of the Society. Professor Klemin of the Guggenheim School of Aeronautics, then a part of New York University, responded to Pendray on 16 March: "I have now had an opportunity of reading your Program of Research rather more carefully," he told him. "It is very good indeed, but I have a few comments to make. . . . I think that neither the Smithsonian Institution nor the United States Weather Bureau should act as sponsors. They should appoint representatives of the Advisory Committee. . . . The estimates of cost are unbalanced. . . . No foundation or individual would authorize an expense of $10,000.00 on salaries and so little on materials, etc. . . . The program is far too general. . . . The actual experiments to be performed are not set forth in sufficient detail. . . ."

Klemin offered his services to Pendray and Manning for a revamping of the proposal but, according to Pendray in an interview by the author, the whole program simply evaporated. Based upon the success of James H. Wyld's regeneratively-cooled motor in 1938, Pendray revived the Foundation idea later that year and took it up with his public relations associate, Morton Savell. James Wyld authored the 11 page prospectus. But again Pendray failed to generate enough interest and once more it died. Slowly, ever so slowly, progress was made nonetheless.

Back in June 1935 a fully operational proving stand was made by John Shesta. The Society also had a new name: at the fourth annual meeting on Friday, 6 April 1934 at the American Museum of Natural History, a motion was passed adopting the new designation, "The American Rocket Society," because many, it was felt, were "repelled by the present name." This was in no way to imply, the Society announced, "that we have abandoned the interplanetary idea." The name was kept throughout the remainder of the life of the Society up until 1963, when it merged with the Institute of the Aerospace Sciences to form the American Institute of Aeronautics and Astronautics. Contrary to the Society's intentions, little in fact was said of interplanetary flight from this point on. The emphasis was placed upon the rockets rather than the spaceship.[158]

With the abandonment of Rocket No. 5 (IAS No. 5, but really ARS No. 1), designed by Pierce, Schachner, and Nathan Carver, work went forward with the test stand. In retrospect, this work turned out to be far more significant than the flights. Much valuable data was gained and led to major developments that were of benefit to the entire country. Shesta's proving stand recorded chamber pressure, fuel pressures, thrust, up to 45 kgs (100 lbs), and exact time of runs. Temperatures were not considered. Thrusts were obtained through a hydraulic plunger while the readings of the gauges were recorded by motion picture. From this data thrust-time curves were drawn. Test engines were clamped down with nozzles aimed skywards, henceforward, one of the standard configurations for test stands. The first series of runs was made at Crestwood, New York, 21 April 1935. Pendray says that all proving fields of the Society were temporary because of the difficulty of obtaining legal permission. Police occasionally threatened to forbid further tests and new locations were often needed. For a while, Crestwood was fairly safe as the tests were conducted in a field adjacent to the Pendray home, though neighbors often complained about the noise. In this first series, five motors were tried, all of them of standard design and none standing up under the intense heat of firing. A second series was conducted 2 June, also at Crestwood. Six runs were made, including one with liquid oxygen and alcohol. One motor had a Nichrome nozzle and another was made of carbon. The Nichrome engine fared well but the other motors burned out. Thrusts varied from 24.5 kgs (54 lbs) to 58 kgs (128 lbs). Burning durations averaged eight seconds.

The third series of static firings took place at Crestwood on 25 August, and included a water-jacketed motor of spun aluminum designed by Willy Ley who had recently emigrated from Germany and had been made an honorary member of the ARS (for Ley's first few weeks in America he was Pendray's guest at his Crestwood home). In one test in this third series of firings a motor exploded scattering fragments over the field. Despite the mishap and the reports of "prolonged explosions" heard in sections of Scarsdale and other towns miles away, a fourth series was conducted 20 October. On this occasion rocket damage was more serious. A woman bystander was injured. She had been struck by a fragment and was rushed to New Rochelle Hospital where an examination showed she had received a compound fracture of the left elbow. Of this regretable incident, *Astronautics* printed not a word. The newspapers, however, made some ado about it. The Society solicited funds from among those present at the test to pay for her hospitalization.[159]

Disaster or not, the 20 October 1935 test was an unrecognized turning point. Among the spectators were, besides the Experimental Committee of Pendray, Shesta, Africano, and others, Dr. George V. Slottman (who still supplied free liquid oxygen); Major Lester D. Gardner, Secretary of the Institute of the Aeronautical Sciences; and Professor Klemin. This represented the beginning of well-placed support or at least overt recognition from a scientific sector that counted, the aeronautical community. Indeed, Gardner and Klemin both joined the Society. The following year the British-born Klemin was to work out the aerodynamics of the non-Society "Greenwood Lake" liquid-fuel mail rocket plane that had been constructed independently by some of the members. Also attending the static tests was a new member, James Hart Wyld, a 23-year-old physics student at Princeton. He was soon to become one of the brightest stars on the Experimental Committee.

Wyld, who had been inspired in the space flight idea by reading Lasser's book, was already undertaking a meticulous study and survey of rocket motors, with emphasis upon searching for a practicable means of cooling. Effective cooling meant long duration runs and would be the key to successful rocket engines that could be adapted for both aeronautical and astronautical applications. The March 1936 issue of *Astronautics* carried member Peter van Dresser's own survey of engines and for the first time depicted Wyld's proposed motor design that passed the fuel through a vaporizing jacket, serving the dual purpose of cooling the motor and preheating the fuel. It was America's first fully regenerative engine, though the Europeans had already developed their own. In the next issue of *Astronautics*, Wyld presented his own article on the problem of rocket fuel feed. At the same time, he aided Shesta, Africano, and van Dresser in building a larger and better test stand. While this project was under way, Pierce and others were flying solid propellant rockets at Pawling, New York, to test aerodynamic shapes (the 4-pound and 6-pound sky rockets came from the Unexcelled Fireworks Company).

That summer too, the Society and Alfred Africano were jointly declared the winners of the 1935 REP-Hirsch Prize for Africano's high-altitude rocket design. This highly acclaimed award signified to the world that the ARS had not only made the year's most outstanding contribution to the science of astronautics but that they were now deemed "professional." The pages of the ARS journal reflected as much. Wyld, for example, was publishing his multi-part "Fundamental Equations of Rocket Motion" and guest articles upon other engineering aspects of rocketry by Eugen Sänger and Alexander Klemin, also appeared.[160]

The summer of 1937 also saw the Society move into new headquarters in New York's Graybar Building at 420 Lexington Avenue. (The following year, however, there was another address change to 50 Church Street.) At the same time there took place an affiliation with other groups so that, in effect ARS branches were created. These were the Yale Rocket Club and the Amateur Research Society of Clifton, New Jersey, which are discussed below. These groups were small and did not last long, but they did portend ARS chapters of the future.

The most important developments of the late 1930s were technical rather than administrative. These were the progress of Wyld's regenerative motor and the completion of ARS Proving Stand No. 2, now on display at the National Air and Space Museum in Washington. Specific details of the Wyld motor are well covered in the literature. By January 1938 he had built his motor and was prepared to test it. It only awaited Stand No. 2, which was ready later that year and made its debut on 22 October at New Rochelle, by test-firing a tubular Monel motor built by Pierce. The new rig proved entirely satisfactory except for one unforeseen drawback. It weighed some 136 kgs (300 pounds) which made it unwieldly to transport. To ease transport, carrying handles were attached and a special trailer for road transportation was devised. Dials provided tank pressure, thrust, and timing data. There was still no facility for measuring temperatures, but many other features made the stand an invaluable research tool. A highly reliable pneumatic system for opening and closing the fuel and oxygen valves and a water-flush system for instantly cooling the motors after test runs were also incorporated. As relatively small engines were to be tested, the capacity for measuring thrusts amounted to only 90.7 kgs (200 pounds), while tank pressures could be read up to 35 kgs/sq cm (500 psi). The nitrogen supply gauge recorded pressures up to 210 kgs/sq cm (3000 psi). The clock was an electric one specially constructed for the purpose, having two hands, one revolving once in 10 seconds and the other once in 100 seconds.

Wyld's motor made its debut at New Rochelle on 10 December 1938, running on liquid oxygen and alcohol. This was the second time the stand was taken out. Three engines were fired that day; first Pierce's, then Wyld's, followed by a tubular regenerative motor submitted by a young Midshipman from the Naval Academy at Annapolis, Robert C. Truax. In the fourth test, the Wyld motor was again tried. The Pierce and Truax engines fired erratically. When first ignited, Wyld's engine refused to register the thrust gauge even though a very large, diffuse, crackling yellow flame shot out. Examination showed that combustion had failed to work back properly inside the motor from the fuse. It was therefore given another try, the fourth run of the day. This time a large yellow flame was produced which shortened into a straight, blue one after a few seconds. The reaction simultaneously rose to 41 kgs (90 pounds) which was steadily maintained for about 13.5 seconds after which it fell quickly as the liquid oxygen gave out. The .9 kg (2 pound) motor was almost cool to the touch. Maximum exhaust velocity was calculated at 2,094 m/sec (6,870 ft/sec) for a maximum thermal efficiency of about 40%. "These figures represent a great advance on those obtained in former tests," ran the report, "and are among the highest ever recorded. The fact that they were reached without severe damage to the motor is especially encouraging and definitely proves the feasibility of the regenerative method of cooling."[161]

Upon entering its tenth year, the Society was filled with feelings of both promise and apprehension. The Wyld engine was clearly recognized as a technological threshold all had been seeking. Now, liquid rockets could be adequately cooled so they could be fired over reasonably long durations instead of prematurely burning out due to overheating. This made the liquid fuel rocket a practicable engine. Yet war clouds loomed upon the horizon and ARS funds were again abysmally low. Wyld himself departed for a position with the National Advisory Committee for Aeronautics at Langley Field, Virginia, which drastically cut the amount of time he could devote to the work. Only the solid-fuel flights at Pawling could be continued up to 1939. Neither the display of ARS Rocket No. 3 in the "Rocketport of Tomorrow" exhibit in the Chrysler Building at the 1939

World's Fair nor some rare ARS talks by Goddard himself were enough to lift the spirits of some. Not until June 1941 was Wyld's motor again tested on the big red stand. That is another chapter beyond the scope of this study. It was, however, the most important technological advance the American Rocket Society was to make. The test runs of 8 June, 22 June, and 1 August 1941, at Midvale, New Jersey, proved conclusively the efficacy of Wyld's design. Just a few days after the Japanese attack on Pearl Harbor, Wyld, Shesta, Pierce, and an electronics engineer with a flair for business, member Lovell Lawrence, united to form Reaction Motors, Inc., of Pompton Plains, New Jersey, for the express purpose of exploiting the Wyld engine for the war effort. The Navy became interested, and subsequently offered an immediate contract for reliable liquid JATO's (Jet-Assisted Take-Off rockets). Out of JATO's came huge, successful power plants for an entire array of projects from the modest Gorgon experimental missile to America's first large-scale sounding rocket, Viking, and to the country's first liquid-fuel rocket propelled plane which first broke the sound barrier, the Bell X-1.

The exigencies of war may have temporarily closed the experimental period of the American Rocket Society, but larger doors were being opened. The American Rocket Society itself not only survived the war, but was born anew. From 1945 it expanded by quantum leaps and became one of the foremost technical societies in the world. Thus, by the time it merged with the Institute of the Aerospace Sciences to become the American Institute of Aeronautics and Astronautics in 1963, it had more than 20,000 members on its rolls. Pendray indeed had reason to be proud of the odd little gathering of fantasy writers in his brownstone apartment that spring evening in 1930.[162]

VII

The British Interplanetary Society

The Liverpool Years

Almost from the very beginning, British interplanetary travel enthusiasts were beset by a number of obstacles. Of these, the most baneful was the Explosives Act of 1875. It forbade all private experimentation or manufacture of "gunpowder, nitro-glycerine, dynamite, gun-cotton, blasting powders, fulminate of mercury, or of other metals, coloured fires, and [solid] . . . rockets . . ." By rigid interpretation, the Act also came to mean liquid-fuel rockets.[163]

The other impediments faced by the British were almost as debilitating. There existed, for example, neither a champion of the cause of the stature of Oberth or Tsiolkovsky. Nor was there an intense public ground swell of interest as found in Germany, Austria, Russia, and America. The great astronautical movement that arose in those countries in the 1920s and 30s barely touched England. The reasons are difficult to fathom. British conservatism and nationalistic retrenchment brought about by the shrinking Empire may account in part for this attitude. But whatever the causes, it was paradoxically British determination which nurtured and kept the Society alive once it was formed, and it evolved into the most viable and prestigious of all the space travel organizations that had germinated in the 1930s and has lasted until this day.

If the English possessed neither the legal nor leadership prerequisites to initiate their own national space travel movement, they did at least benefit from a sufficiently strong science fiction base from which to develop. In this they had much in common with the Americans. In fact there was an "American connection" in the formation of the British Interplanetary Society which came via Hugo Gernsback's *Science Wonder Stories.*

Phillip Ellaby Cleator, a 23-year old structural engineer from Wallasey, Cheshire, on the south bank of the Mersey River near Liverpool, was an ardent science fiction fan who was to have his own story "Martian Madness" published in *Science Wonder* for March 1934. Cleator may have obtained his copies of the American magazine from F. W. Woolworth's for three pence each. Later BIS member and science fiction writer Arthur C. Clarke recalls that bundles of *Science Wonder* and other pulps arrived in England as cheap and plentiful ballast in the bilges of ships and then resold to Woolworth's. In any case, Cleator first learned of the American Interplanetary Society (and presumably the VfR too) from an item in "The Reader Speaks" column in the April 1931 issue of *Wonder Stories.* It was an appeal by the Society's Secretary C. P. Mason for new membership as well as a report on the successful turnout for Robert Esnault-Pelterie at the American Museum of Natural History. Cleator was prompted to write to Mason on 10 August for further details: "Quite by accident," he wrote, "I came across a short description of the American Interplanetary Society in a magazine from your country. On behalf of the Science Research Syndicate, I shall be greatly obliged to receive from you full particulars together with the necessary membership forms."

Secretary Mason could not doubt that his British correspondent was fully sincere in pursuing the interplanetary problem as Cleator's imposing stationary from The Scientific Research Syndicate included on the left-hand margin the Syndicate's business: "Chemical, Physical and Electrical Researches—Analysis and Synthetic Compounds—Experimental Apparatus Manufactured for Research Work—Radio and Television Devices—Inventions Perfected." Cleator was the "Director of Research" and his home at 34, Oarside Drive, Wallasey, Cheshire, was the Syndicate's headquarters. Cleator had in fact inherited his father's engineering business and was also continually engaged in his own scientific pursuits. He had always been fascinated with science and even as a boy of 14, had made crude experiments with a "pistol rocket." The roots of his awakening to space travel, he remembers, stemmed not from literature but from the movies. In the 1920s, he says, "I happened to see a documentary film concerning the unusual properties of radium, which concluded by predicting that such a source of energy might one day be used to send a spaceship to the Moon." Perhaps this film was *All Aboard for the Moon,* produced by Bray Studios of New York in 1924, and later released in Europe. *All Aboard* was indeed a documentary and one that included a spaceship to the Moon propelled by radium.[164]

As the international space travel movement developed, particularly from the early 30s, Cleator's interest deepened. He read everything he could on the subject and even began correspondence with Hugo Gernsback, not long after the formation of the American Interplanetary Society. It is a shame that almost all of Gernsback's early correspondence was thrown out and that Cleator's files and library were destroyed during World War II by a direct bomb hit on his home. It is well-known, however, that one reason Cleator was led to form the British Interplanetary Society was the result of an article he read in the *Liverpool Echo* in August 1933, containing a brief statement by W. A. Conrad of the United States Naval Academy on the possibility of reaching the Moon by rocket. Shortly after, on 8 September 1933, the *Echo* printed Cleator's response: "It is significant to note that Mr. Conrad is an American," he wrote. " . . . England is years behind. So far as I know, the problems of interplanetary travel have received little or no real attention here as yet. . . . Before me as I write is a letter I have recently received from C. P. Mason [this letter has not been found]. . . . He says, ' . . . feeling that there is a field for a British (Interplanetary) Society, which we hope to see organized at the earliest moment.' . . . The immediate formation of a British Interplanetary Society is imperative, if we are to keep pace with other progressive nations. . . . Perhaps those interested will communicate with me?"[165]

The British public remained unmoved. "Now the *Liverpool Echo* is a well-known and widely read paper," Cleator recalled, "and the appeal appeared under a caption calculated to attract attention. But it did not. Public apathy in the matter, it became evident, was not to be shattered easily. It seemed that nothing would suffice to disturb public indifference. But Mr. [N.E.] Moore Raymond, special correspondent for *The Daily Express,* thought otherwise. Not only would he place the Society on its feet, but he would become a member. And so the impossible happened. How he persuaded the editor of *The Daily Express* to spread the news across the front page, I do not know. It is sufficient that he did. And so my troubles ended—or should I say really began? At any rate, from the deluge of correspondence which immediately resulted, there emerged the British Interplanetary Society."

As with the American Interplanetary Society, and probably with the VfR and Russian groups, these were actually preliminary gatherings prior to the inaugural meeting. One was held in Cleator's home on 25 September 1933 in which half a dozen or so enthusiasts showed up and were shown a rocket motor made by their host. In his laboratory Cleator also demonstrated the combustability of fulminate of mercury. Amongst those present was 19-year old Leslie J. Johnson, a clerk with the Liverpool Education Offices who had earlier founded Liverpool's first science fiction club (which soon became defunct), "The Universal Science Circle." Johnson also brought a former member of his group, Colin Henry L. Askham, who was also a radio ham.

The official founding date of the British Interplanetary Society was Friday, 13 October 1933. The address was an accountant's office—Room 15, 2nd Floor of 81 Dale Street, Liverpool 2. The offices belonged to a parent of one of those attending, Herbert Chester Binns, Cleator's boyhood friend. The founding members numbered 15, including two German citizens who were not present but were made "Honorary Founder Fellows." In Cleator's account in his *Rockets Through Space* (1936), he says he started with six people and that by towards the end of December the membership climbed to 15.

These members were all regarded as "Founder Fellows." Besides Cleator, who was the obvious choice as the Society's president, they included: Colin Askham who became the Society's first Vice-President; Leslie Johnson who was designated the Secretary because he owned a typewriter, obsolete though it was; Miss A. C. Heaton, a pharmacist, the only lady present, and the designated Treasurer; Herbert C. Binns who provided "a convenient meeting place" and who became the Society's accountant; Percival Norman Weedall, the Society's first Librarian; J. Toolan, an automobile engineer; and Messrs. Thomas McNab, James A. Free, and E. Roberts, "young but enthusiastic members." Comparable to the founders of the American Interplanetary Society, just about all of them were avid science fiction fans, science fiction then going hand-in-glove with the space travel movement. There were also the "Founder Fellows," the father and son chemists Richard and Raymond Thiele, respectively, of Cologne, Germany, friends of Cleator's; and the "Honorary Founder Fellow" R. S. Chambers of *Chamber's Journal* who afforded the group generous publicity. Cleator called them all "the nucleus of the British movement in rocketry."

By January 1934 the first issue of the *Journal of the British Interplanetary* appeared, bearing Cleator's Wallasey address as BIS headquarters. The magazine's cover was adorned with a futuristic black and white silhouette rendering of a rocketship flying by a block of skyscrapers and an oversized Moon. It was the winning design selected by popular vote in a contest suggested by Cleator—with himself claiming the one guinea prize. Subscription for members was set at £2 2s or two guineas per annum which could be paid quarterly "by arrangement." Associate membership, for those under the age of 21, was 5 s. per annum. This was surely a reflection of the youthfulness of the founders. Ordinary meetings were held fortnightly. Upon its debut that same year, the science fiction tabloid *Scoops* also printed news of the BIS, and by reciprocal agreement was advertized in the BIS *Journal*. Later meetings of the Society were held in the upstairs room of a cafe in Liverpool's Whitechapel. It was an unpretentious place which exactly suited the needs of the Society. Cleator says "it was centrally located; it remained open until late hours; it provided light refreshments as and when required; and its charges were absurdly cheap."[166]

Phil Cleator's first order of business was, besides generating as much publicity as possible, flying to Germany to confer with VfR officials. In January 1934, three months after the formation of the Society, he met Willy Ley in Berlin and, like G. Edward Pendray before him, was accorded a statesman-like tour of *Raketenflugplatz*. By that time, however, both Germany and the VfR were upon hard times so that little could be shown. The leading experimenters—Klaus Riedel, Hans Hüter, and the aristocratic young Wernher von Braun—were gone. They had begun working for the *Reichsheer,* the German Army. In fact, the VfR had collapsed by the time of Cleator's arrival, and all he recalls seeing was "a collection of dilapidated buildings." The remaining experimenters were at loose ends after the failure of Rudolf Nebel's Project Magdeburg. Still, Cleator considered his mission "very fruitful."

During his ten-day sojourn in Germany, Cleator spent two days with Ley. (Cleator says he also met Rudolf Nebel "and found him polite but uncommunicative, as befitted a knight of the New Order. So I returned to Ley, and we dined pleasantly together, and discussed plans for the future.") In addition, Cleator may have also spoken with the young journalist Werner Brügel. Brügel, the author-editor planned an IRKA (*Internationale Raketenfahrt-Kartei* or, International Rocket Travel Information Bureau) which was publicized in the BIS Journal. But sadly, in Germany he attracted the attention of the Gestapo instead of altruistic space travel buffs and was jailed for two years. (Another account attributes his incarceration in a concentration camp to the earlier publication of his book.) Through the courtesies of Willy Ley, Cleator was also "able to obtain introductions to many of the leading experimenters throughout the world. The result was very gratifying, many well-known rocket experts having since joined the British Society [i.e., by April 1934]."

No meetings are mentioned other than with Ley and Brügel, so that the "introductions" must have meant obtaining addresses. Ley himself was made a BIS Fellow, followed by Perelman and Rynin of Russia, Esnault-Pelterie of France, von Pirquet of Austria and Pendray of the United States. Later this distinguished roster was further augmented by: Ernst Loebell, founder of the Cleveland Rocket Society; Harry Grindell-Matthews, inventor and rocket experimenter; Wally Gillings, British science fiction editor; Count and Countess von Zeppelin, son and daughter-in-law of the inventor of the Zeppelin; Dr. Otto Steinitz; Friedrich Schmiedl, the Austrian rocket mail pioneer; Olaf Stapledon, the science fiction writer; and Professor A. M. Low, the British pioneer in the development of radio-controlled guided missiles in World War I and long an advocate of space flight.

The 18 January 1934 issue of the *Daily Express* disclosed another effect of the Cleator trip. From the vantage point of half a century it makes amusing reading, especially in light of Cleator's promotion of flight into space at near astronomical speeds. "He arrived home in Wallasey on Tuesday," the paper said, "but was so exhausted by a 12-hour airplane flight in a gale that he went straight to bed and slept the clock around."[167]

Following the cementing of German and other European contacts, and an abortive attempt to duplicate Pendray's "brainwave" by showing *Frau im Mond,* Cleator's next order of business was experimentation. In actuality, some work had already been performed before he had departed for the continent. An almost cryptic report was carried in the first number of the BIS *Journal:* "Among the various projects on which the Society is working at the moment is the construction of a rocket car. Except for a few technical details in connection with the motors themselves, the plans are now complete. It is hoped to gain much valuable information from the experiment." Almost nothing is said thereafter in the *Journal* of this curious project, but scant details may be added from other sources.

In addition to reporting upon his trip, the 18 January 1934 *Daily Express* also commented upon the Cleator rocket car. The Germans, they noted, perhaps facetiously, had tried all sorts of propellants for their rocket automobiles, including beer! Cleator preferred a more sober fuel. He was said to have considered a liquid oxy-acetylene combination. The vehicle was planned for a 564 km/hr (350 mph) speed. (In a letter to the author, Cleator says that speeds of at least 160 km/hr or 100 mph, were anticipated and that the length of the car was planned at 3.6 meters or

12 feet.) "It would be best," he wrote at the time, "to have a rocket car specially built, but the Society cannot afford this. I am therefore going to make a car in my own laboratory in Wallasey from parts of old motor cars."

From an item in *The Autocar* for 2 March 1934, the car's purpose is defined. It was to function as a mobile test bed for both solid and liquid propellant devices. It was intended first to develop "a really serviceable 'motor' before making any attempt to justify their [the BIS's] magnificent title." In the second issue of the *Journal,* Cleator somewhat obliquely hinted, without giving specifics, that while the press had been highly favorable and beneficial to his cause on the whole, the remarks made by *Autocar* concerning the project were not entirely accurate. Cleator noted also the published assertion in another paper that Mars might be annexed to the British (or other) Empire in the not too distant future was an unqualified "exaggeration." At all events, the Cleator car was never built.

One reason for the abandonment and inability to start new projects was the perennial problem of money shortage. "No one is more eager than I am to organize and to begin our share of actual experimental work," Cleator told his members. "But I have not become resigned to the fact that we cannot hope to do this until membership is greatly increased, or until we receive financial aid from some outside sources." From his new-found American friend, G. Edward Pendray, with whom he was to engage in years of correspondence before actually meeting face-to-face in the early 1970s, he elicited a long and detailed letter explaining to the BIS readership the need and the feasibility of undertaking experiments and that ARS rocket Number 1, complete with stand, had cost a grand total of $49.40. Cleator ran it as an article entitled "Why Not Shoot Rockets?" but he still could not raise the necessary funds. The Explosives Act of 1875 was yet to be encountered, though Cleator, at least, knew where the Authorities stood in the matter of financing rocket research. The Air Ministry had evidenced not the slightest concern. While in a letter received from the Undersecretary of State, it was stated that "we follow with interest any work that is being done in other countries on jet propulsion, but scientific investigation into the possibilities has given no indication that this method can be a serious competitor to the airscrew-engine combination. We do not consider that we should be justified in spending any time or money on it ourselves."[168]

In the meanwhile, another controversy presented itself, from a quite unexpected quarter. It came from within the thin ranks of the BIS itself. In October 1934, Member J. G. Strong reflected upon the American Interplanetary Society name-change to the American Rocket Society. "We should do well to follow their example as soon as possible," he concluded. Cleator, in consultation with the BIS Council did not take the suggestion lightly. He staunchly opposed the change, charging a pandering to public opinion. It also seemed to him "that a change in name regardless of the reason for it, would be universally misconstrued as an admission of doubt, as a confession that the interplanetary idea only belongs to the realm of extravagant fiction." His motion was upheld.[169]

If the British Interplanetary Society could not afford to fire rockets, perhaps a world-wide amalgamation and consequent pooling of resources was the answer. Cleator early favored internationalization but as close to this goal as anyone came was the customary exchange of literature. Anglo-American relations were always strong. And from the near moribund *E. V. Fortschrittliche Verkehrstechnik,* successor to the VfR, free subscriptions to their journal *Das Neue Fahzeug* were available to all BIS members by 1936. Members were likewise entitled to the American Rocket Society's *Astronautics.* Earlier, in the Spring of 1934, efforts had also been made to utilize the radio equipment and talents of Vice-President Askham and member James Davies to communicate on either the 40 meter (7,074 kilocycle) or 20 meter (14,148 kilocycle) short wave bands with other members of the world's rocket societies. This grand plan never reached the air waves. Two years later, however, Cleator happily enlisted the services of Ralph Stranger, Secretary of the World-Wide Radio Research League (WRRL). Through WRRL's organ, *Science Review,* the BIS message was broadcast even further. Stranger was duly accorded an honorary Fellowship. Soon after his nomination he published an account, both in the *Journal of the British Interplanetary Society* and *World Radio,* of the "three-year-old mystery" of radio disturbances from outer space as discovered by the American physicist Dr. Karl G. Jansky of the Bell Telephone Laboratories.

Two other examples of early BIS aims to internationalize space travel were the Society's acceptance of Pendray's proposal to establish a common fund for the publication of a universal journal. This really meant a union of the three leading English-speaking groups, the ARS, BIS, and the new Cleveland Rocket Society, discussed below. The BIS would have been required to contribute £50 per annum. By February 1935 it was evident that this was unworkable, and the scheme was "reluctantly abandoned on account of cost." A more bizarre means of bringing kindred souls together was the BIS or at least Secretary Johnson's espousal of Ido, the International language.[170]

The most direct way to raise the necessary capital for research was to follow the American example, by establishing an experimental fund. Cleator announced this move in his editorial for May 1935. The BIS Secretary-Treasurer, then L. J. Johnson, was placed in charge of the Research Fund" and an Experimental Committee was also formed. By 1937 financial conditions were still in an unsatisfactory state, and it was reported by a new section, the Technical Committee, that the Research Fund remained at "microscopic proportions." So were the experiments. Drawing from the £5 Research Fund, "rough tests" of from 80 to 120 suggested fuels were made, "the majority of them eliminated as useless." The chemical portions necessarily had to be small because of the expense involved. Two of the tests alone amounted to £2/10s. o.d. Apart from these unpretentious efforts that were part of the celebrated BIS spaceship project discussed later, practially nothing is known of actual BIS experimentation during the 1930s or at any other time.[171]

By the 1936—1937 period the exasperating Explosives Act of 1875 was fully disclosed and from there on any tests that might have been undertaken were undertaken clandestinely. Eric Burgess, for example, recalls the surrepitious construction of a liquid oxygen-gasoline rocket of member Alan E. Crawford during the late 30s. It was a virtual carbon copy of the old German two-stick Repulsor but with one interesting innovation. For protection, the motor was coated with a metallic spray. It is not known if the 1.2 meter (4 foot) rocket was ever fired.[172]

Of far greater consequence to the history and indeed the future of the BIS were the initial liquid propellant investigations of Ralph Morris, a London member. Early in 1936, in response to Cleator's appeal for experimentation, Morris came forward and offered to make a liquid propellant rocket himself. In a letter to G. Edward Pendray, dated 10 March 1936, Cleator sums up the rest: "He sent his plans etc., to me and between us we evolved a fairly decent design. He promised to let me have a full report in due course, and meanwhile set about the task of construction. The next I heard from him was a long letter to say that his experiment had been prohibited by the Government!"

Cleator had approached at least two MPs, one of whom he had met on a train. Both were negative. After a long delay a reply was received in August in the name of the Secretary of State that in some unexplained manner liquid fuel experimentation would contravene the Explosives Act of 1875, otherwise called the Guy Fawkes Law. Cleator recalled his complaints against the odious injunction years later in his article "Autopsia" in the BIS Journal for May 1948.[173]

This Governmental posture was all the more irksome since just a year after the founding of the BIS, Cleator, in company with Professor A. M. Low and several other members of the Society journeyed to London early in May (1934) to the Apex International Air Post Exhibition to meet Gerhard Zucker, the young German rocket mail pioneer who was being allowed to launch several experiments throughout England. Cleator and his friends spoke with the German and only saw a stationary aluminum rocket but on June 6 the first of several widely publicized trial firings was held. The postal authorities would not officially sanction the dispatch of letters from Sussex Downs, near Rottingdean, but before four newspapermen two rockets were fired anyway—with 2,864 pieces of mail aboard which were afterwards recovered and postmarked at the Brighton Post Office. This was England's first rocket mail. The reasons why Zucker was able to fire rockets amidst a great fanfare of publicity—and do it repeatedly—in spite of the "anti-rocket" law which had so restricted the British, was because the Postal authorities were apparently unaware of the rocket ban. The prospect of possible commercialization for the benefit of the British populace may have also temporarily blinded the Secretary of State and other authorities. At any rate, Zucker was able to easily get away with illegal rocket flights whereas BIS members could not.[174]

It was however, well known that the Englishman Harry Grindell-Matthews was conducting his "secret" rocketry experiments at his large estate at Mynydd-y-Gwair, Wales. Yet the authorities did not bother him either. This must have been particularly galling to Cleator as Cleator had virtually introduced him to rockets. Matthews, moreover, was so immune from prosecution that he felt secure enough to have private roads put through his estate and also have obstructing telephone poles removed. Trespassers were kept out by six feet high barbed wire fences and other means. Years later Matthews' relationship with the Government was revealed. Apparently he had signed a contract to produce an aerial torpedo that could bring down a Zeppelin, or any aircraft. He is said to have received £25,000 in Government subsidies. By the time war clouds were gathering in earnest upon the horizon, much of the Matthews fortune had dried up. Skilled labor was also difficult to keep because of the demands of the British rearmament program. The year was 1938, and out of impatience, anxiety, and the threatening geo-political situation, the secretive inventor decided to lift the veils. Not all of the details would be disclosed—just enough he hoped to lure the support he desperately needed when his work was nearing fruition. At the same time he hoped to interest the Admiralty in a new submarine detector. The mysterious torpedo turned out to be a so-called snare rocket. In 1938 it was frightening enough and probably itself added to the war scare rather than comfort people. Steadied by gyroscopic controls and stabilizer fins, the missile would entangle the propellers of attacking enemy airplanes by discharging a series of parachutes dangling enormous lengths of steel wire.[175]

In Scotland, the "Paisley Rocketeers," discussed below, also escaped the Guy Fawkes Law in the 1930s, as did Alwyn (later, Sir Alwyn) Crow. But Crow's work came under the Ministry of Supply and also was afterwards carried out in Jamaica. The Manchester Interplanetary Society illegally fired rockets but they were not so lucky. Their story can not first be told without relating the relocation of the BIS from Liverpool to London, the stepping down of P. E. Cleator, and the end of "the Liverpool Era."[176]

The move can be explained in two ways. The usual explanation, and certainly a valid one, is that by 1936 the greater proportion of BIS members resided not in Liverpool but at the capital, in London. Consequently, there was also a shift of influence. This culminated in a general agreement that the BIS could be better run in London. Cleator gracefully resigned and completed the process. The second explanation is that at the same time—or actually a short time before—there had been a mounting resentment within Cleator's own ranks at Liverpool against his long-standing domination of BIS affairs. J. Happian Edwards introduced the plan to open a London office that appeared in print in the June 1936 issue of the *Journal* under the title, "A London Section?" The Council was particularly moved by his rationale that the Society would be in a better situation to enlist an even more substantial following if a London branch were so formed. Edwards also observed that the Society's stationery looked more "Amazing Storyish" than "like a serious society," and recommended that a new design be procured. Both suggestions were adopted. Edwards was quickly named the official London representative and could be reached at the 362 Radio Valve Company, Ltd., of which he was an executive. By this time also, Professor A. M. Low had been named Vice-President for London. Simultaneously, C. H. L. Askham, the original Vice-President, still held that post at headquarters in Liverpool. By November the trend towards a shift of power and location to London was irreversible. Of all the internal events, Cleator had no control. Bowing to this fate, he promised to the Society that it "would still have his every support."[177]

The first meeting of the London Branch took place on 27 October 1936 at Professor A. M. Low's offices, 8 Waterloo Place, Picadilly. Low was named President. K. W. Chapman and Miss Elizabeth Huggett (later, Mrs. J. H. Edwards) were nominated the Joint Honorary Secretary. Arthur C. Clarke, a 19-year old Somerset farmer's son who had first encountered the idea of space travel through David Lasser's book, was designated the Honorary Treasurer. Director of Research went to J. Happian Edwards. Edward J. Carnell, a leader in the British science fiction movement, was named "Director of Publicity." Professor Low who read the Proceedings, "gave an extremely interesting address drawing a comparison of prejudiced public opinion throughout the ages with the aims of the Society today, and emphasized the importance of not being discouraged by this, or in our turn, becoming prejudiced." J. Happian Edwards then gave a short address. All contemporary science fiction authors, he said, "put the ultimate time for space travel at the year 2000, but he was firmly convinced that it would become a fact before that date."[178]

A session of the London Branch was called on November 15 at the Mason's Arms pub on Maddox Street, not far from Picadilly—the Mason's Arms became a favorite rendezvous for the BIS thereafter and was affectionately re-named "The Space-Shippers' Arms." At that meeting the London Constitution Committee was formed which would work to overhaul the constitution.

The most important meeting at the Mason's Arms took place on Sunday evening, 7 February 1937, and the two outstanding matters discussed and subsequently adopted were: (1) that the BIS officially transfer to London; and (2) that a Constitution be drawn up by the Special Committee of the London Branch. Even before the move was promulgated, the February 1937 issue of the *Journal* declared in an editorial that "We on Merseyside have such confidence in our London Branch that we have not the slightest doubt the transfer, if it is effected, will be a great step towards a much larger and more active Society." The first London issue of the *Journal* came out months later, in December 1937, from the new headquarters address at 92 Larkswood Road, South Chingford, London E.4—the home of H. E. Ross. Professor A. M. Low was the new president, while the two vice-presidents were P. E. Cleator and L. J. Johnson.[179]

The London Years and the Manchester Interplanetary Society

President Low's historical connection with the BIS went back to the Society's foundations. When Cleator visited the Germans he also obtained from Willy Ley, besides the introductions to some of the leading European astronautical pioneers, a list of all those Britons who had been in touch with the VfR. Low was the most prominent. Low was at once conferred with an Honorary Fellowship and soon became the rallying point for the London group. That Low's professorial credentials were somewhat questionable if closely examined (he was an honorary professor of the Royal Artillery College) is inconsequential, considering his unflagging support of both the space travel cause in England and the BIS. He publicized the BIS as much as he could, for example in his *Armchair Science,* and in 1938 promoted, albeit unsuccessfully, a nationwide rocket-plane contest to interest youth; response was nil. It was also Low whom Willy Ley first met after leaving Hitler's Germany. From London, early in 1935, Ley then went to Liverpool where he stayed for a week at the Wallasey home of a friend of P. E. Cleator. From thence he traveled to America, where, like a roving ambassador he next stayed with the Pendrays.

Cleator's role was far from forgotten upon Low's accession to the BIS Presidency. In 1936 his *Rockets Through Space* came off the presses. As the first British book on the subject it did much to boost the movement for years to come both in England and other countries. As an immediate consequence of *Rockets,* the BBC invited Cleator to present a short address on the subject. A motion picture was also released that same year (1936) which likewise infused a new-found enthusiasm amongst the populace, Alexander Korda's production of *Things to Come,* based upon the novel by H. G. Wells. The film became a classic overnight. Its utopian vision of a century hence—the year 2036—after Everytown had been bombed in 1940 and then resurrected by science and reason, was Wells' personal credo: "For man no rest and no ending," says the son of the first astronaut about to be shot to the Moon. "He must go on, conquest after conquest. First this little planet with its winds and ways, then the planets about him, and at last out across immensity to the stars." The BIS could find no fault with these leaping sentiments. It praised the film but expressed one objection. Associate member D. W. F. Mayer of Leeds criticized the space gun. "A little mathematics soon indicates the absurdity of this scheme," he began. "Mr. Wells has incorporated into the Space-Gun scenes of the film an idea which no astronaut has seriously considered since the days of Jules Verne. If the 'Man in the Street' is to be introduced to the possibility of space travel via the medium of films—especially films with as much publicity as was given to *Things to Come*—it is up to the writers of them to make sure their facts are reasonably accurate, and not give the public the idea that modern astronautical societies resembling the Baltimore Gun Club [the fictional but probably the world's first society for advocating and indeed later carrying out a flight to the moon, in Jules Verne's *From the Earth to the Moon,* 1865]. Play the game, Mr. Wells!"[180]

Almost buried beneath Mayer's critical review of *Things to Come* in that February 1937 issue of the *Journal,* was an almost plaintive cry: "An Associate Member, aged 19, wishes to correspond with members of his own age. Please write to: W. Heeley, 25, Crayford Road, Manchester, 10." Heeley was but one of the under-21 years-of-age members of the Manchester Interplanetary Society that had been created on June 1st of the previous year by Eric Burgess. President Burgess was 16. This was the average for all five members, including the two girls, Madeline Davis of Staveley, Darbyshire, and Lillian Dawber, Burgess' cousin from Manchester. "Fair haired" Burgess, as one paper called him, had really begun his space career earlier in life. Similarly to P. E. Cleator, he undertook some crude rocketry play at age 13 or 14. The concept of spaceflight came to him through Edgar Rice Burroughs' Mars novels. In time, letters were sent to Leslie Johnson in Liverpool for advice. Comparable to Cleator, Burgess expanded his club through the power of the press. The father of Founder Fellow Trevor Cusack (who joined November 19) was a newspaperman and first suggested this approach. Once a story was generated in a paper, even a local one, like the *Ashton Under Lyne Reporter,* it could soon be picked up by a larger national daily paper, as the *Daily Express,* or a big city paper, as the *Manchester Evening News,* and so on. The theory indeed worked in practice. The launching of the mimeographed MIS journal *Astronaut* in April 1937 also helped matters, as did a lecture or two by Burgess before the Manchester Astronautical Society in the Central Library. Although typical of the British astronautical movement in the 30s, membership never grew by leaps and bounds and totaled to no more than 40 or 50 people by 1939 when it was then known as the Manchester Astronautical Association. The name-change was, it might be said, a repercussion of the MIS's first experiments.[181]

By March 1937 the MIS boasted of 16 or 17 in their ranks, with meetings generally held in Burgess's house at 683, Ashton New Road, Clayton, near Manchester. Out of their meager pocket-money (Burgess, like the others, was still in school attending the Municipal High School of Commerce), several small powder rockets were crudely fabricated and hundreds of small tests of stability and cellular arrangements made. To Burgess and his dedicated young members, these simple skyrockets were far from play. He hoped to instill in himself and his followers "practical experience" with the desire successively to progress towards liquids. By October the contemplated Rocket MIS-1 had already been drawn up. It was to carry a litre of liquid oxygen and gasoline (petrol). Burgess had gone out and purchased the litre can. Compressed nitrogen was to force the mixture into the combustion chamber following the standard American pattern. But for the moment, only powder for sky rockets was available. On 27 March 1937 about 100 people gathered at the Dingle, Clayton Vale, to see Burgess and his fellow MIS members demonstrate six rocket launchings. Five of the six barely left the ground, and the sixth exploded upon ignition, hurling metal in all directions which injured some of the spectators. A plain-clothes policeman promptly closed the event.[182]

Later Burgess and three companions were summoned to the Manchester City Police Court on charges that they had violated clauses of the Explosives Act of 1875 as well as an 1890 Order in Counsel. The case appeared as Rex (The Crown) vs. the Interplanetary Society of Manchester. Quite naturally it made the headlines for the duration of the proceedings, which the sagacious Burgess fully exploited. Fortunately for his side, Kenneth Burke, who was a friend of a *Daily Express* reporter, was himself caught with the space flight dream and served without compensation as Burgess's barrister. Fortunately also, Burgess was both plucky and knowledgeable about the theory, if not yet skillful enough about the construction of rockets; he simply "dazzled the court with science," according to press reports. The prosecution, which included Government explosives experts, were admonished by the magistrate for not being aware of the background of rocket technicalities. The Stipendary Magistrate, J. Wellesley Orr, was also a fair man. The result was that the summonses were withdrawn upon the promise that neither Burgess nor any other member of the Manchester Interplanetary Society ever use potassium chlorate and sulphur again. That the chlorate had been used as an ingredient in the ignition powder rather than as the propellant was immaterial, for the decision at the Manchester Police Court effectively shut out all possibility of future experiments by the MIS and the BIS too. In London and Liverpool the BIS paid the closest attention. Any hope of the nullification of the rocket clause in the Guy Fawkes Law was entirely dashed. In the following year, Cleator angrily penned a four-page sarcastic essay on "The Rocket Ban" in the Manchester Society's own journal, *The Astronaut.*[183]

A final postscript should be added to the MIS story. The upshot was that a heated disagreement arose between the proponents of experiments and those who maintained that only a private journal was feasible "to raise and spread interplanetary ideas amongst the rising generation. . . . " A schism occurred with the confirmed "experimenters," Burgess and Cusack, forming a new society in December 1937 called the Manchester Astronautical Association. The remaining Manchester Interplanetary Society was essentially run by William Heeley and Harry E. Turner, the latter an insatiable collector of science fiction literature and an employee of the Imperial Chemical Industries.

In 1938 the MIS itself helped fill out the ranks of the BIS by affiliating with it, but voluntarily disbanded upon the commencement of World War II. As for Burgess and Cusack, the Magistrate did not stop them and they simply continued to experiment in secrecy, at least one of these tests being conducted under the cover of a foggy day at their station in the moors. Theoretical work also continued with a design for a liquid-fuel sounding rocket capable of ascending 12.2 kms (40,000 feet) completed by the end of 1940 which Burgess offered unsuccessfully to the British Government for antiaircraft use. With the collapse of the BIS and MIS because of the war, the

Manchester Astronautical Association continued (though Cusack was drowned in 1941 when the ship upon which he was serving as wireless operator in the Mercantile Marine was sunk by enemy action). It had amalgamated with other groups mentioned below and grew to an organization of about 200; under the leadership of Burgess and Ken Gatland it was the only wartime astronautical society in Britain.

At the beginning of the war Burgess had unsuccessfully attempted to interest the military in rocket development. An interview with a Navy research official in London ended with: "no interest in liquid-propellant work."[184]

The final chapter of the pre-war BIS—for it was reborn after peace was declared—was the famous BIS spaceship. The project is well documented, not only in the *Journal of the British Interplanetary Society* (issues from December 1937 to July 1939), but in *Flight* for 12 February 1942; *Spaceflight* for February 1969 and March 1974; and in the Smithsonian Institution's *Annals of Flight,* No. 10.[185]

Faced by the frustrations of a phantom "Experimental Fund," and a seemingly insurmountable law that prevented any tangible experimentation, the stalwarts of the BIS's Technical Committee under the direction of J. Happian Edwards began slowly to map out the design of a feasible compound solid propellant spaceship capable of flying to the Moon. Much of the preliminary laying out of this project was done in Arthur Janser's comfortably appointed rooms on Great Ortmond Street.

The BIS spaceship blueprint had been completed at Summer's end in 1939 just before the war broke out, and the BIS as a whole put away its drafting boards but not its dreams. Each detail of the craft had been painstakingly designed by authorities in his field. Members of the Edwards Committee were: H. Bramhill (draftsman); Arthur C. Clarke (amateur astronomer); Arthur V. Cleaver (aircraft engineer who became one of Britain's leading rocket engineers); Maurice K. Hanson (mathematician and a leading figure in British science fiction circles); Arthur Janser (chemist from Austria); S. Klimantaske (biologist); and Ralph A. Smith (turbine engineer and the BIS' famous space artist). Technical assistance was also given from time to time by Richard Cox Abel, the engineer J. G. Strong, and C. S. Cowper-Essex.

Perhaps inspiration was also provided by a young American midshipman, Robert C. Truax, who visited the Society in July 1938. Truax, then on a training cruise on the battleship *USS Wyoming* to Denmark, Sweden, England, France, and Spain, had made previous arrangements with BIS officials to take part of his one-week leave in London to visit the Society and present a talk on his own rocketry investigations. Truax recalls that BIS officialdom was then thoroughly emersed in their huge solid-propellant lunar ship, though he himself was strictly a liquid-propellant man. Nonetheless, "they were certainly open to my suggestion of experimentation," Truax says, "but they really could not experiment because of their financial situation as well as the lack of facilities." The BIS delighted in Truax's regeneratively-cooled chambers that had already modestly proven itself with a compressed air-gasoline combination for a thrust of about 55 kgs (25 pounds)—making it one of the first liquid fuel rocket engines ever seen in England; this motor was later tested on the American Rocket Society's test stand on 10 December 1938. Mostly, Truax continues, he found the BIS in a fever pitch about their Moonship. They were even excitedly discussing who would be the likely passengers.

Truax, in his low-keyed, unassuming manner, becalmed them somewhat and urged them to devote more of their precious time to practicalities than needlessly deliberate about who would become the first British astronaut. The advice was heeded. Squadron Leader D. Ross Shore provided voluntary and professional help in the design of parachutes. John W. Campbell, Jr., afterwards editor of *Astounding Science-Fiction* and considered the "father of modern science fiction" in the post-Gernsback era, recommended a gyro stabilizer that was subsequently adopted. Harry E. Ross helped work out the firing controls. Ross, who is also the principal chronicler of the BIS spaceship story, credits much of the overall design to Edwards and Smith "who had been close friends and interested in the possibility of space travel since schooldays. In fact the idea of cellular-step construction was Edwards' and the engineering embodiment Smith's."[186]

Work began from casual meetings of the Technical Committee before it was officially formed in February 1937. There was no overall plan, but there existed a considerable body of open literature, much of it based upon experiments performed elsewhere. From February 1937, a more formal program began to be developed. Based upon their specialties, "teams" were asked to initiate what they could and then to help fit all the pieces into a coherent whole. Dipping into the ridiculously small Research Fund, the Austrian research chemist, Janser, in collaboration with Edwards made a survey of from 80 to 120 propellant combinations by the end of the year. These apparently amounted to no more than "pinch" and burn tests, though pastes were also tried. Those that looked promising were slated for proper combustion runs in "actual propulsion tubes as soon as we can afford the tests." They never could. Smith designed a test stand but it was never built and in fact was criticized. Textbooks in hand, Janser and Edwards moved from carbon dioxide to organic combinations, then to colloids with metal additives. The latter were the most favored for their high caloric value and relative exhaust velocities. The highest was determined to be boron and oxygen. In the final analysis, Janser and Edwards concluded that, contrary to prevailing opinion, solids contained as much energy potential as liquids and that the main difficulty with the latter was constructional. Cooling and fuel supply (via high compression pumps) were overwhelming engineering challenges for a large scale rocket and the highest theoretical liquid combination, mono atomic hydrogen and tri atomic oxygen (ozone) were not found "in a sufficiently stable and safe form yet." Unbeknown to them, von Braun's German team of hundreds of well-financed top-flight engineers and chemists were at that moment perfecting the high compression, light weight pumps the British could hardly imagine. From the cost standpoint, then, compact solid units were early selected but with smaller, throttable hydrogen peroxide liquid or steam systems serving as torque and other control jets. The solid propellant was never properly [187]

Closely following the German and American experiments, especially the contemporary American Rocket Society proving stand tests, it was determined that nichrome nozzles "would cost an amount quite beyond our present resources." But with optimum design of a combustion chamber following Eugen Sänger's formulae, higher chamber pressures would be possible and incomplete combustion negated. To confirm some of this data, crude miniature nozzles were constructed from which gases (nitrogen, hydrogen, carbon dioxide, and steam) under various conditions were expelled. "Unluckily," reads a parenthetical remark in the Janser-Smith report, "tests were limited to these gases which one of our members was able to obtain for nothing." In regard to the space ship itself, cost-effective considerations evolved it into a multi-step cellular configuration. Costs were determined from Cleator's *Rockets Through Space* and Oberth's studies. It was estimated that a cellular rocket making a one-way trip to the Moon (presumably an unmanned mission) would cost £20,000. A cellular rocket (manned) designed for a two-way voyage was expected to cost £200,000. The frugal-minded Britons effectively argued that cellular construction with solid-fuel rockets would drastically reduce the expense of space flight over large liquid-fuel space-craft such as those designed by Oberth. This prompted the BIS to project optimistically a lunar voyage at the price of a single destroyer. Expenses would also be slashed in that the solid-propellant "cellules" were mass produced and expendable. Each cellule was asbestos-bound, making the system fairly safe as well as cheap. Each rocket could also be ignited individually if necessary so that a certain amount of thrust control was possible.

Altogether 2,490 cellules made up the ship which consisted of six steps arranged in tiers of diminishing-sized rockets towards the top. Overall dimensions were 32 by 6 meters (105 by 19.7 ft). Total weight was 900,000 kgs (1,980,000 lbs) of which the propellant comprised 1,150 tons (2,300,000 lbs or 1,043,280 kgs); the pressure cabin for a crew of two or three, 0.75 tons (1,500 lbs or 680 kgs); batteries for ignition and other power needs, food stores for 20 days, tools, water and air, 5 tons (10,000 lbs or 4,536 kgs); and the general structure—Janser had considered lightweight plastics where possible—275 tons (550,000 lbs or 249,480 kgs).[188]

Mathematician Maurice K. Hansen and amateur astronomer Arthur C. Clarke worked out navigational computations, while at the start of 1939 the others filled in the myriad details of the life-support system with its air-locks, air-conditioning plant, food stores, and control systems. (The stepper switch control system appeared to derive from Clarke's association with the telephone arm of the British General Post Office, according to Eric Burgess.) Radar was then under secrecy wraps, though it was known that short wave radio would suffice as a possible means of communication across space. Transit navigation was principally by the centuries-old optical observation of the planets and stars, but with a new twist. This ingenious device designed by Smith was known as the Coelostat.

The Coelostat was actually built and is the most famous of the pieces of "hardware" to come out of the BIS space ship study. Smith published a complete account of it in the July 1939 issue of the BIS *Journal,* although it was first publicly displayed before the Society at a meeting on 7 March in the Science Museum in South Kensington, London. Briefly, it was meant for navigation, to allow Coelostat observations from a space ship that revolved (at a rate of three revolutions/second) in order to simulate gravity for the occupants. Two mirrors, placed at 90° to one another, revolved together and two more mirrors formed a stationary periscope into which the observer looked. The light falling into the rotating mirrors thus passed to the stationary mirrors as everything slowly revolved. During main thrust periods, however, light-weight plasticized inertial instruments such as an altimeter, speedometer, impulse meter, chronometer, gyroscope, and accelerometer were to do the navigation job automatically. Work on the development of a suitable altimeter and speedometer was actually begun, the former starting out as alarm clocks. This problem proved to be more troublesome than first appeared. After dismantling five time-pieces to learn their mysteries, a spring-weight combination and aluminum disc fly wheel arrangement was tried. But this too presented problems and as Ross says, "the sands of peace ran out before it could be completed." A high-energy lightweight primary battery was also investigated by the electrical engineer Ross, but had to be abandoned due to lack of time and money. Most fascinating of all was the anticipated payload to be carried, the details of which were also revealed by Hanson in January.

The BIS planned mission to the Moon was as it actually happened 30 years later in July 1969, for "scientific observation and mineral prospecting." Details down to razor-blades were considered, along with onboard exercises, high protein foods, electrically-heated ovens, aluminum or plastic utensils, balsa wood pencils, light-weight playing cards, rubberized yet heat-resistant space suits and puncture repair kits, signal rockets for use on the moon's surface, dark goggles and sunburn lotion, dynamite charges for removing large lunar rocks, specimen tubes and reagents, a fairly powerful telescope, a microscope, a spring balance and gravity pendulum, a cine-camera and ordinary miniature camera, flat-bottomed shoes, a tent, and "an adequate supply of various kinds of paper money with which the intrepid explorers will pay for their return to civilization should they land in one of the more barbarous regions of the earth." The very last issue of the pre-war *Journal* focused upon one of the most difficult problems, that of landing the ship both on the Moon and the Earth. For the solution the call went out for "any suggestions re the design of shock absorbers. . . ." Who can say what further proposals would have come about had not the war intervened? The BIS abruptly ceased activities late in 1939 after an extraordinary general meeting.[189]

Cleator wrote prophetically to Pendray on 14 September 1939—the very week he had planned to come to the United States to see his friend in person. He had visited Grindell-Matthews a few weeks earlier he said, and "Matthews was still trying to interest the Powers That Be, so far without success. For myself, I have given that up—though it would certainly be ironical if a world war were required to demonstrate the potentialities of the rocket. Somehow, I hate to think of the device I have fathered for so long turned to destructive purposes . . ." The drama of the BIS final days is also found in the pages of *Astronautics* of the American Rocket Society. In the November 1939 issue came the last BIS correspondence to the United States before the war. Arthur C. Clarke wrote to Alfred Africano, then the President of the ARS: "Owing to the War, it will be impossible for the BIS to carry on the active existence and arrangements have been made to put it in cold storage for the duration. In order that our work will not be entirely lost, whatever happens to us, I am sending you a few copies of our last two 'journals.' . . . If the worst comes to the worst—which I don't for a moment think it will—we hope that the ARS will be able to see that our work has not been entirely wasted." The BIS was reborn again after the conflict and on 20 July 1969, member Arthur C. Clarke, who had survived the war, spoke before millions of television viewers around the world as Apollo 11 touched the surface of the Moon. He could say with full conviction that he and his fellow BIS members had already been there before.[190]

VIII

The Other Societies

The Cleveland Rocket Society

By the early 1930s there was no question that enthusiasm for space travel was an international phenomena, in short, an international movement. Not only did it take the form of societies or clubs. Sometimes there were also publicity and money-seeking or misguided individuals such as shadowy Dr. Darwin Lyon, or "Professor" Robert Condit, allegedly of Ohio University. The 1928 Sunday supplements delighted in showing Condit about to prepare for an "anti-gravity" spring-rocket trip to Venus, his giant rocketship poised in his garage and Condit in his tee-shirt. But the Professor claimed he was prevented from meeting his scheduled astronautical rendez-vous for two reasons: his "backers" refused to allot him further funds and, a fierce meteor shower blocked his path to Venus.

However, there were also many valid expressions of the belief in interplanetary travel, such as individual but more solidly scientific experimenters and small alliances of fellow enthusiasts who sought to emulate the larger groups. It may be said that they too, in varying degrees, further helped arouse among the public an awareness of the new but as yet untried science of astro-nautics. The earliest and most active of these smaller organizations was the Cleveland Rocket Society.[191]

The CRS was largely the brain-child of a dynamic 31-year-old German-born engineer, Ernst Loebell, who had already been inspired in rocketry and space travel in his native country. While never a member of the VfR, he closely followed all the latest developments in German rocketry and the space travel movement. Loebell also engaged in "critical yet enjoyable discussions" with Karl Poggensee at the *Ingenieur Akademie* of the University of Oldenburg (later known as the *Hindenburg-Polytechnikum*). On 13 March 1931 Poggensee successfully launched a solid-propellant rocket carrying a radio transmitter, an altimeter, camera, and velocity meter to an altitude of 457.2 meters (1,500 feet), perhaps making it the world's first successful sounding rocket. Both Loebell and Poggensee learned from each other, especially as they shared another interest: amateur radio.[192]

The real turning point for Loebell occurred just after he graduated as a mechanical engineer from the University of Breslau in 1927. He landed his first job—in the Engineering Department of the Berlin-Witenau branch of the American-owned Otis Elevator Company. In February or March of 1929 they sent him to their headquarters in Yonkers, New York. He later moved to Cleveland which had a sizeable German community and a fine engineering society. In January 1933 the Cleveland Engineering Society requested him to present a lecture on rocketry, if possible with a model for their Technicraft Show. Loebell eagerly consented. He was also able to construct a bullet-shaped chrome-nickel-steel replica of the VfR configuration. One day, he said, it would be possible to deliver mail from Berlin to New York in 30 minutes with such a ship. One of the visitors of the show was Edward L. Hanna, licensed airplane pilot, sportsman, and grandson of the famous railroad-steamship-coal-iron-millionaire Marcus Alonzo Hanna, "President-maker" and erstwhile political boss of Ohio during the McKinley era. This chance meeting led to many others, and it was not long before Loebell and Hanna had formed a friendship based upon a mutual desire to bring the rocket into being. Within a month the idea developed into a strong conviction that they could interest others. Thus, the Cleveland Rocket Society was born. From the start, Hanna's influence in the creation of the Society was both providential and substantial (initially at least) so that he may rightly be called a co-founder. CRS Headquarters were in Room 410 of the Hanna Building.[193]

Support also came from elsewhere. Soon after his first meeting with Loebell, said Hanna in his recollections, "an engineer by the name of Charles St. Clair [of the Domestic Fuel Co.] came to us and offered both time and effort. His financial contributions made possible the purchase of equipment, and a nucleus was formed from which the [CRS] Technical Department grew."

Loebell also attracted German-born Dr. Hugo K. Polt, professor of Germanic languages at Western Reserve University (now Case Western Reserve), who became CRS's second president. The attorney John Crist also joined and was to participate in many tests besides offering his legal services. Harold Carr and Fred Donley were particularly helpful; both were machinists and helped construct the first four motors. Donley, a civil engineer, was the CRS Vice-President. Cecile Shap-iro had access to a printing shop and was able to publish several of the five issues of the CRS's journal, *Space,* which appeared between December 1933 and September 1934. Loebell planned to distribute it to 165 cities in the United States and also to the other English-speaking nations. Originally, the magazine was to be named *Rocketry.* Perhaps this title should have been retained as neither Loebell nor the CRS were space-oriented. Their position was made emphatic in the August 1934 issue of *Space.* In his article "Attaining the Escape Velocity," CRS Secretary Charles A. Prindle, Jr. says: "The Cleveland Rocket Society is concerned solely with the perfection of the rocket motor for terrestrial [sic] transportation. The detailed study of interplanetary flight will come years hence when rockets have proved their worth to the commercial world on this planet."[194]

By the Summer of 1934 CRS membership amounted to 50. A Technical Staff had been formed that was divided into four independent departments: Mechanical Engineering, Electrical Engineering, Radio Engineering, and Theoretical Engineering. As member Karl Spangenberg held a Master of Science degree in electrical engineering, and because of Loebell's long-standing interest in the subject, radio played an important part in CRS thinking about flight rockets. Spangenberg himself wrote several articles on his specialty for *Space,* including "Ultra-Short Wave Antennae For Plotting of Rocket Trajectories."[195]

The most illustrious of CRS members was Lieutenant Commander Thomas G. W. Settle, then the world's champion stratosphere flyer. His balloon "Century of Progress," rigged at the Goodyear-Zeppelin Airship Dock at Akron, Ohio, ascended to 18,665 meters (51,236 feet) on 20 November 1933.[196]

Informality marked the organization with no regularly meeting scheduled. In spite of this drawback, the CRS was able to elicit considerable attention through local newspapers, radio stations, and occasional small exhibits. Although the CRS journal, *Space,* was far less effective as a propaganda tool for the Society and the movement than were the ARS and BIS publications. Barely enough copies were printed for distribution in Cleveland, let alone the 165 cities and foreign countries Loebell had in mind.[197]

The first issue of *Space,* moreover, ran a story on page three which, in the scientific world, temporarily spoiled the reputation of the rocket movement. It announced that a *manned* rocket had ascended to six miles! This was the notorious Fischer brothers hoax. Bruno Fischer was supposed to have been the constructor of the rocket, and Otto, his brother, the pilot. They were alleged to have taken off 4 November 1933 from the island of Rugen, in the Baltic off the German coast. The hoax was perpetrated not in Germany but in England; it first appeared in the sensationalistic London tabloid *Sunday Referee* for 5 November 1933. To compound the fraud, the writer for *Space* (apparently Hanna) innocently announced that "A photograph of this amazing rocket is at the Society's headquarters where members may view it. The cost of the ship was about $35,000 [,] all of it being raised from private companies interested in the experiments." The photo the CRS proudly owned was not of the fictitious Fischer rocket but of Nebel's Magdeburg Pilot vehicle superimposed upon an appropriate background. It should be noted that the American Rocket Society, the British Interplanetary Society, the Austrian Society for High Altitude Exploration, and MosGIRD and LenGIRD were similarly "taken in." On the positive side, so far as the CRS was concerned, the CRS journal *Space* still accomplished much in inspiring and bringing in new members.[198]

The CRS mimeographed bulletin, *Space,* fulfilled the role accomplished by similar bulletins of the ARS; reporting details of tests. But these were only static tests. For all the activity, no CRS flights were ever made. A large plumbing tube test-stand was erected at a "Proving Field Laboratory" at the rear of Hanna's Lake Erie waterfront estate; it was situated on Waite Hill, Kirtland, about a dozen miles east of Cleveland. Six liquid oxygen-gasoline engines were built and test fired here.[199]

"First, we will perfect our small rocket motor," said Loebell, "then we will apply this motor to an airplane and prove that that airplane will go higher and much faster than any previous planes. We will fly this first plane minus a pilot, steered by radio. When that has successfully reached and returned from the stratosphere we will build a large plane and equip it with a rocket motor, and let a pilot go along."[200]

The first test was made at sundown, 21 October 1933, with an egg-shaped motor weighing about 1.4 kgs (3 pounds) and 15 cms (six inches) long. Ignition was by blowtorch. The motor performed admirably, even though a shortage of pressuring liquid nitrogen necessitated Loebell's resorting to liquid oxygen for forcing in both the fuel and oxidizers, especially as it was a cold day. Tascher says the egg-shaped motor test fired nine times with average durations lasting ten to 20 seconds. (Charles Prindle's report in *Space,* December 1933, says 120 seconds.) Calculated thrusts for most of the CRS motors was about 14 kgs (30 lbs). In the second test, one week later, 28 October, a smooth run was brought to an abrupt end by the bursting of the gasoline tank. "During the entire scene," the official report reads, "Mr. Loebell displayed coolness and courage. The last piece of the shattered tank had not landed before he had opened all valves relieving the pressure within the oxygen tank." At Loebell's request the explosion was not reported in the press. He explained that it would be bad publicity, that it could hurt member recruiting and that it might even get the CRS into trouble with the law (promoting a dangerous activity or something like that). The press was sympathetic and complied.

The third test on 26 December 1933 was the most spectacular of all, especially as it was made in a blinding snow storm and stinging wind. Nitrogen was now available and electrical ignition (i.e., a spark plug) introduced. Prindle and one or more other young members took credit for this improvement; prior to this time the blow-torch sufficed to light the engine. Yellow flames leapt out of the exhaust nozzle, then smoothed into bluish-white accompanied by a healthy, steady roar. "It's a success," Loebell proclaimed after the run. Loebell then set about drawing plans for the construction of a larger, .6 meter (two foot) long, .3 meter (one foot) diameter motor, "with power enough to lift 454 kgs (1,000 pounds) as long as the fuel lasts."

By December the CRS boasted two score of members. This was still not enough to increase the Society's treasury to continue its tests. Perhaps one of the newer members, Alfred F. Stern, had the answer. Stern was also an avid stamp collector, a member of the United States Philatelic Society and later a professional stamp dealer. Following the examples of several enterprising other rocketeers, he suggested the printing of special rocket post cachets and selling them for profit. Loebell agreed. Stern, named "Cachet Director," had 100 made, each bearing the legends "First Flight", and "Cleveland Rocket Society" in the upper left hand corners. An advertisement was also published in leading stamp magazines such as the *Airpost Journal* for May 1934, offering the covers for sale at $1.00 each. Loebell was pleased with the result: "These envelopes sold like hot cakes, and gave us the necessary funds for our last motor." The ironic twist is that, in Loebell's words, "we never did launch a completed rocket." Even today there must be several collectors of rocket post who still believe they possess genuine examples of rocket mail.[201]

On Saturday, 16 June 1934, Movietone News representatives gathered on the Hanna estate to document the CRS fourth test, but the thrust needle barely moved. Then as a result of the 28 October 1933 explosion, Carl Hanna, Edward's father, forbade further experimentation on his farm. The CRS was forced to find another site. This was the Rau farm in Highland Heights between Richmond and Bishop Roads, the property of a retired lawyer friend of Ed Hanna. Rau permitted Loebell to cut trees and make a clearing.

A much larger test stand was erected and the motors were bigger, but the tests proved unsuccessful. Nonetheless, the last CRS motors, built in 1935, were remarkable. In essence, Loebell made a regeneratively cooled system that predated James H. Wyld's regenerative cooled rocket motor by three years. The Loebell engines, of which two were made, were never fired, however, because CRS research funds had dried up. Copper tube coils surrounded the combustion chambers and carried both fuel and oxidizer which were then injected in the normal manner and then ignited. Loebell's motors were also very large for their day. They were .5 m (18 inches) long, weighed 9 kgs (20 lbs), and were calculated to deliver thrusts between 2,268-3,175 kgs (5 to 7,000 lbs). CRS member John W. Burke later acquired the motors and in 1937 prepared one for a test, but again, money ran out. During a World War II scrap drive one of the motors was destroyed while the surviving engine was donated to the Case Institute of Technology in Cleveland.[202]

The value of the CRS work, however praiseworthy, was highly questionable from the scientific standpoint. Almost no statistical data was recorded. Loebell, the engineer, should have known better. Still, the CRS framework stand and (23-foot) 7 meter long solid roof covered observation and control trench was considered by the BIS to have been "one of the best-equipped rocket testing grounds in existence . . . the satisfactory completion of the necessary tests will lead inevitably to the flight of their first rocket-driven projectile." This was never to be.[203]

Loebell turned to both private industry and the military for assistance but without success, and during those difficult Depression years of the mid-1930s he was having difficulty finding employment. The CRS, accordingly fell on hard times, and spirits sagged particularly low when the Great Lakes Exposition to be held in the Summer of 1935 at Cleveland turned down an exhibit which the Society had offered to give without charge.[204]

In 1937 the situation seemed reversed. In April of that year the French Government's *Minis-*

tére du Commerce et de L'Industrie requested the CRS to exhibit a rocket at the Paris International Exposition in the imposing halls of the *Palais de la Découverte.* Loebell was bouyant, and he immediately sent his acceptance. He also asked the French if they would be willing to pay for the shipping charges of a model of a proposed long-distance mail rocket and one of the two liquid-cooled (regenerative) chambers. The French agreed. The long-distance rocket was one quarter-scale and stood about 2 meters (6 feet) tall and weighed 159 kgs (350 lbs) loaded with fuel. Like the 1929 *Frau im Mond* spaceship, it opened on hinges to reveal its interior. Within it were, besides the motor and its ancillary equipment, a parachute and spring-activated timing device for releasing the parachute, a compartment for holding mail and a short-wave receiver. The magnesium alloy rocket was designed for a 40-48 km (25-30 mile) altitude, with a speed "close to sound" and a thrust of up to 680 kgs (1,500 lbs). To add to the realism of the exhibit, postage stamps that had already flown via the mail rockets of the Austrian Friedrich Schmiedl were displayed. The *Bulletin de la Société Astronomique de France* and the *Journal of the British Interplanetary Society* wrote favorable reviews of the show and exposition organizer Robert Lencement sent Loebell two copies of a *Diplôme Commémoratif* (Commémorative Certificate) in grateful remembrance of his part in the Exposition; one of the certificates was awarded to Loebell, the other to the Society. Earlier, Lencement proudly informed Loebell that his model rocket was placed "in the best place of the Astronautical room" and that "it raised the greatest interest of our *President de la République* [Albert Lebrun] during the visit he paid last Wednesday [21 July 1937], and I have given him plenty of informations [sic] about it and the Cleveland Rocket Society."

Charles A. Lindbergh was also impressed with the exhibit during one of his several trips to the Continent at this time, according to Loebell. Lindbergh was already an ardent believer in rockets. It was directly through his intercession that Robert H. Goddard received considerable financial support for his own rocket work by the Daniel and Florence Guggenheim Foundation for the Promotion of Aeronautics. We know of no efforts on the part of Lindbergh to obtain comparable support for Loebell.

Following the Paris International Exposition of 1937, Loebell was again broke. The accolades for his exhibit were very satisfying but brought neither him nor the CRS additional revenue nor promise of benefactors. Further, Loebell lacked resources with which to test the duplicate regenerative motor he had so gladly sent to Paris. By the time the Exposition opened, the CRS lacked sufficient manpower to maintain the Hanna "Proving Field," or "Rocket Aerodrome" as Loebell liked to call it. For all practical purposes, the CRS ceased to exist by the Summer of 1937.[205]

Still, it was about this time that Loebell began corresponding with Hans K. Kaiser of the *Gesellschaft für Weltraumforschung* (Society for Space Travel) of Berlin, and later Cologne, Germany. The correspondence lasted until almost the eve of World War II. Kaiser wrote to Loebell on 2 August 1939 with a hopeful request that he donate whatever CRS models of rockets and drawings remained so that they might be displayed at an anticipated World's Fair in Rome and an International Transportation Exhibit in Cologne, scheduled for 1940. But it was far too late for both Loebell and Kaiser. An exhibit on par with the 1937 show in Paris or the 1939 World's Fair in New York City was no longer possible with the opening of hostilities.

In 1938 Loebell became associated with William P. Lear later of Learjet fame and a man whose own capacity for invention was phenomenal. During his lifetime Lear took out more than 150 patents on his inventions including the car radio, the eight-track stereo for cars and the airplane directional finder. Lear hired Loebell as a stress analysis engineer in 1941, Loebell remaining with the company for 26 years until his mandatory retirement at age 65 in 1968; even at that date he was still giving talks on rocketry and spaceflight. He died in Cleveland in 1979. CRS co-founder Hanna also remained captivated by the rocket idea but had long ceased to be active in the field. In the 1930s he reputedly set aside a trust fund of $20,000 for his burial on the Moon, in case neither he nor his wife would see Moon flight. This pretentious wish was not fulfilled. Hanna died 16 November 1968, eight months before Apollo II touched down on Tranquility Base.[206]

GALCIT, PRA, Yale Rocket Club, et al.

In 1936 a small pool of scientists and other enthusiasts at the Guggenheim Aeronautical Laboratory of California Institute of Technology, under the titular head of Dr. Theodore von Kármán, the world renowned aerodynamicist, initiated the GALCIT Rocket Research Project.

Their work culminated in the creation of the Jet Propulsion Laboratory and several great American aerospace achievements, notably the United States' first large-scale liquid fuel sounding rocket, the WAC Corporal; the casting of solid-fuel propellants, a break-through making very large scale solid propellants possible; and the first American Jet-Assisted-Takeoff (JATO) units, facilitating aircraft take-offs. Although GALCIT is included in Cedric Giles' list of rocket societies, the GALCIT founders did not set out to form nor ever call themselves a society or club; they did generate scientific papers but really for internal or self-educational use only, and were not concerned with proselytizing their ideas to the general public. In fact, they had little regard for the societies or popularization of rocketry and space flight. (Though it is noted that the driving force behind the GALCIT team, Dr. Frank Malina, paid some attention to the ARS during GALCIT's earliest years and also had the results of some of his progress published in the ARS *Journal*.) The GALCIT Rocket Research Project (1936—1938) is thus aptly named and is not dealt with in this study. For those interested in the GALCIT story, this group is well covered in the literature cited below.[207]

There were several other smaller American groups which barely subsisted and quickly fell by the wayside. Some were one-man affairs. All lacked money. Few, if any, bulletins were issued. Hence, these groups are largely consigned to obscurity. Thankfully in 1944 the editor of *Astronautics,* Cedric Giles, compiled a list of all the world's known rocketry and astronautical societies so that those who shared the same dreams and ideals of the larger and comparatively wealthier societies would not be totally forgotten.[208]

The earliest of the lesser American astronautical or rocketry organizations that began before 1940 was the Peoria Rocket Association (PRA), founded at Peoria, Illinois, on 27 March 1934 by Ted S. Cunningham. Amateurish by any standards, the PRA nevertheless represents a gauge of how deeply the astronautical and rocketry movement had penetrated by the early 1930s: to small town America. The motto of the PRA, "Organized to Promote Faster Transportation," faintly resembles the purpose of the contemporary German Registered Society for Progress in Traffic Technics (EVFV). In fact, Cunningham was in contact with the later German group, Hans K. Kaiser's *Gesellschaft für Weltraumforschung,* and may well have been inspired to start his own society by learning of the earlier EVFV. As the PRA's motto also seemed to imply, rockets were only one means of achieving the desired "faster transportation." Cunningham, a waiter in his father's restaurant, constructed a 58 cm (23-inch) long miniature Zeppelin. The full-sized version was to be propelled by compressed hydrogen and ignited [!] by a 2760° C (5,000° F) electric arc. Cunningham also claimed to have sent an instrumented balloon to an altitude of 14 kms (nine miles). A rare copy of Volume I, No. 1 of the *Journal of the Peoria Rocket Association* of January 1939, states: "On 27 March 1934 the Peoria Rocket Association was born. A small experimental balloon to be sent aloft to determine conditions in the higher strata of air was launched. Disaster followed. But through many, many struggles the desired information was found. Many members gave up the idea and this led to the collapse of the Association the Spring of 1935. After the collapse Mr. Cunningham continued studying alone. In the spring of the present year [actually 1938] a new idea was conceived by Mr. Cunningham and a new Association was planned. On 30 November 1938 the first meeting was held. Five members make up the new Association."[209]

Undoubtedly Ted Cunningham and his five members were sincere. But it is equally obvious that they lacked genuine technical awareness of rocketry and space flight and had little chance to attract fellow Peorians or anyone else with real technical and organizational acumen they needed. Giles reported that the PRA *Journal* survived for only four issues, from January to April 1939. Very probably, the PRA experienced a second, and fatal collapse before the year's end.[210]

Soon after Ted Cunningham's first, short-lived PRA was formed, came the Yale Rocket Club of New Haven, Connecticut. It was begun in late 1935 in Room 339, Wright Hall, Yale University, by four students of the school and one outside individual. The founders were: 20-year old Franklin M. Gates, Electrical Engineering (Class of 1938); 21-year old Merritt A. Williamson, Metallurgical Engineering (Class of 1938); 19-year old John A. Beattie, Physics (Class of 1939); Beattie's room mate 21-year old Francis Morse, Mechanical Engineering (Class of 1939); and Vincent Anazski, a New Haven machinist in his 30s. Anazski was a man of considerable practical experience, and though not a Yale student like the others, shared the same youthful enthusiasm.

Philosophically and technically, the small Yale Rocket Club showed great promise in promoting and working for the cause of rocketry and space travel. Three of the original five founders were already members of the American Rocket Society. These were Gates, Beattie, and Anazski.

Gates brought them together and first learned their names through correspondence with Peter Van Dresser, editor of the ARS Journal *Astronautics.* From these obscure beginnings the Yale Rocket Club expanded to almost 40 people on the mailing list who wished to receive notices of the Club's meetings. Gates and Williamson were the most active members, Gates serving as Chairman for three years until his graduation, and Williamson as Secretary-Treasurer for two years. Afterwards he assumed the Presidency for two years until 1940 when he completed postgraduate work at Yale.[211]

One of the first acts promulgated by Chairman Gates was an expansion of the Club by proposing an amalgamation with The American Rocket Society. Negotiations for this arrangement dragged out for over a year. Through personal talks with ARS President Alfred Africano and with approval of the ARS Board of Directors, the Yale Rocket Club finally achieved the satisfaction of being granted Charter No. 1 as an affiliate or branch on 1 January 1937. Affiliation meant discount membership, entree into the ARS, and occasional loans of ARS spokesmen and artifacts for talks and displays at Yale. The *Yale Alumni Magazine* for 4 February 1938, for example, reported that: "Jules Verne may have had definite ideas about a trip to the moon in a rocket, but they were as impracticable as they were untrue, according to Alfred Africano, president of the American Rocket Society, who spoke to the Yale Rocket Club last week. Man's only hope of getting off the earth is to wait for the invention of processes to release atomic energy." The enterprising founder and first president of the Yale Club also sought out Robert H. Goddard. He wrote to Goddard on 14 April 1937 requesting scientific material for use at their meetings, but Goddard politely declined, saying typically that his experiments and research were not yet completed.[212]

Merritt Williamson, the Secretary and second President of the Club, characterizes the Yale Rocket Club's early existence as "struggling." "I say 'struggling,' " he writes, "because not only were the late thirties far from being a period of affluence, but also rockets were hardly an acceptable topic of conversation. Our fellow students treated us with tolerant good humor and the majority of professors and at least one Dean thought we would be much better advised to devote our attentions to important matters! In the midst of a very discouraging response on the part of the faculty, one bright spot was the attitude of Dr. Charles H. Warren, Dean of the Sheffield Scientific School, who provided us with a small office at 702 Sterling Tower which the Club maintained until 1940."

Franklin Gates recalls a similar picture of indifference and an occasional bright spot on the part of the Yale faculty. Upon its formation, he says, the Yale Rocket Club hoped for financial support from the University but quickly realized that this was not possible. Gates and his fellow members were grateful enough for the one or two professors who did show their support, one of whom, Al Conrad, Dean of the Electrical Engineering Department, provided some technical advice; support also came from Professors B. R. Teare and L. C. Lichty. Classrooms were also made available for the Club's regular meetings, the blackboards being handy for formulae and diagrams.

Mainly the Yale Rocket Club's activities were centered on the endless but stimulating theoretical discussions of escape velocities, trajectories to the Moon, space ship designs, optimum performance propellants, and super metals for making the spaceship. Thwarted by an almost total lack of funds, the Yale Rocket Club conducted only minor experiments but planned many more; rocket motors were made but never fired. The job of educating the public or "arousing interest," as Williamson phrases it, proved to be their key function. They also educated the ARS. Williamson, now a professor of engineering management at Vanderbilt University at Nashville, Tennessee, translated a number of German papers on rocketry for the Society. He thus made information of active interest available to ARS experimenters. "Articles by Hans Grimm, Guido von Pirquet, and Eugen Sänger among others," says Williamson, "found their way into English." One of the more significant of these translations was Eugen Sänger's *Neure Ergbnisse der Raketenflugtechnik,* appearing in the ARS journal *Astronautics* October 1936 issue as "The Rocket Combustion Motor." Gates and Williamson also collaborated on an article, "Astronautics, A New Science," published in the Winter 1937 issue of *The Yale Scientific Magazine,* also appearing in abstracted form in *Science Digest* for May 1937. Encouraged by the interest shown in this article, Williamson published another one in the Summer 1938 issue of *The Yale Scientific Magazine,* entitled "Research Problem in Rocket Engineering." Other ambitious, albeit abortive attempts in contributing towards the understanding of rocketry and space flight were a *Hand-*

book of Rocketry started by Gates and Williamson, and a work tentatively titled *Thermodynamics of the Rocket Motor,* by Williamson.

The members of the Club tried hard to acquaint the world around them with the potential of rockets. The Yale University Library was encouraged to open a rocketry and astronautical section by buying all available foreign and domestic books in the field. Lectures were also arranged for New Haven audiences. Even with Franklin Gates' matriculation in 1938, along with several other members, Williamson still kept up his prodigious activities. He kept the Club going until 1940, when he completed post-graduate work at Yale. "After I left," Williamson says, "there was simply nobody around to continue the work and in any event the war came. The Yale Rocket Club ceased by 1940."[213]

In a reflective article in the *Yale Scientific Magazine* for March 1965, Meritt Williamson asks a fundamental question that may apply to all of the rocket societies discussed in this book: "Did our little group at Yale influence the progress of rocketry?" Williamson answers his own question: "Although some might be dissatisfied with the proof offered, my answer is an unqualified, 'Yes.' From the mailing list of 38 persons who, between 1937 and 1940, were interested enough in rockets to want to be notified of meetings and who attended when time permitted, fifteen are now listed in the tenth edition of *American Men of Science.* Of these listed, six are known to the writer to have been involved in rocket research and development. At least three others who were not listed are known to have worked with rockets. One of the early members is, at present, a high official in a company devoted to the design, development and production of rockets."

Several of these individuals may be identified. Williamson himself did not end his rocketry activities in 1940. During his Navy service from 1944 to 1946 he was made a Guided Missile Project Officer assigned to the Navy's new rocket center at China Lake, California, where he worked on the Tiny Tim and Project Bumblebee programs. Williamson also claims to have been one of the first official students of jet propulsion, having formally studied the subject at the California Institute of Technology's GALCIT. Another founder of the Yale Rocket Club, John A. Beattie, after receiving his M.S. in Physics at Yale in 1939, transferred to the Massachusetts Institute of Technology where he completed another Masters in Physics in 1940, writing his thesis on rocket propulsion based upon results obtained from a gaseous hydrogen-air rocket motor mounted in a static test stand. The thesis led to a real rocket job; helping Aerial Products Inc. of Merrick, Long Island, develop black powder rocket flares for use on blimps to chase submarines for the Navy. More directly involved with the Space Age is former Yale Rocket Club member Dr. Edward H. Seymour, at one time General Manager of Reaction Motors, Inc., and presently an administrator with the ARS's successor, the American Institute of Aeronautics and Astronautics (AIAA). The story of the Yale Club Williamson concludes, "may serve as an inspiration to others who have the urge to pioneer and as a warning to those who tend to belittle or discourage initiative wherever it may exist however crazy or futile it may appear to be in the light of present day knowledge."[214]

A more bizarre group in the 1930s than the serious minded Yale Rocket Club was the International Cosmos Science Club. The ICSC does not appear on Giles' list as it was primarily a science fiction group. As a reincarnation of the old Scienceers of New York, interest in interplanetary flight was maintained and rocketry experiments performed, after a style, by the President of the New York Branch, William S. Sykora. Sykora, a 22-year-old tool designer at the Westinghouse X-Ray Company, fancied himself already scientifically-bent, his "science-hobbyist" articles often appearing in the mimeographed pages of the ICSC's *International Observer.* The NYBICSC (New York Branch of the ICSC), as Sykora pretentiously preferred to call it, amounted to perhaps 30 or 40 members and was founded in 1935. Rocketry experiments were conducted the same year, the findings duly reported in the *Observer.*[215]

The first series of tests went off 10 March 1935, but only one of the four rockets worked. A second series of tests also failed. And so it went until the fourth series which was held 22 September 1935, and hailed—quite incorrectly—as "the first successful rocket air mail flight made in the United States." (The *Ellington-Zwisler Rocket Mail Catalog* cites a flight made 4 May 1904 by the balloonist C. C. Phelps of McConnellsville, New York, though unsubstantiated, and *several* rocket flights from Struthers, Ohio from 1 July 1931 to 30 April 1932). Sykora's attempt, made from the Holmes Airport at Astoria, New York, experienced one accident. A boy spectator was injured by a piece of the aluminum rocket which imbedded in the muscle of his left arm. The *New York American* for 23 September sensationalized the incident saying that the rocket carried a "secret explosive" and that Sykora expected the rocket to go straight up in the air for two miles.[216]

Neither William S. Sykora nor any other member of the NYBICSC ever again experimented with flight rockets, although Sykora did attempt, with the help of his father, to make "a miniature liquid-fuel rocket motor."

Another amateur but more sober group, also started in 1935, was the Junior Rocket Society of Crown Point, Indiana. Its founder, Warren E. Pierce, wrote to Robert H. Goddard on 19 February 1935 for advice as he could learn little from his high school science teacher. Goddard was encouraging to the young man, but otherwise offered him little tangible to go on. Goddard recommended his *Method of Reaching Extreme Altitudes* from the *Smithsonian Miscellaneous Collections* "which may be found in any large public library," and he also suggested an article he had written in the *Scientific American.*[217]

The Westchester Rocket Club of Westchester, New York, founded in 1936, was a decidedly more professional organization dedicated to the perfection of the rocket—and ultimately towards the realization of interplanetary travel. The moving spirit was Nick Limber whose name often cropped up in the model aircraft magazines of the period. Limber, like Franklin Gates of the Yale Rocket Club, was an active member of the American Rocket Society and sought to use this connection to best advantage. In 1938 he appears as both an officer and "reporter" on the ARS staff, and was largely responsible for its supplemental *Bulletins to Astronautics.* By that period also, Limber could report that his Club was headquartered at 1462 Leland Avenue in the Bronx, and that his members had already been hard at work on a theoretical study of landing gear as well as the construction of a set of 3 meter (ten foot) autogyro wings. Limber and his fellow engineering student members of the WRS, Charles H. Grant, E. Brygider, A. Moskowitz, and Kurt Fisher, made it their specialty to arrive at a foolproof means of recapturing lost rockets, and also to provide stability without the use of gyroscopes. In 1937, Tucker Gouglemann, Secretary of the Club and a junior at Columbia University, also designed a mercury tube parachute release. This was followed up by laboratory tests at New Rochelle, and a test flight for this and the autogyro devices was scheduled for May 1938. No known launch was made, though Limber does seem to have designed and flown his own powder rockets. Because of the seriousness of these young experimenters and their relative proximity to the ARS, the larger society granted affiliation (No. 2 after the Yale Club), and also gave permission to use the ARS test stand when it was complete. But little if anything was actually carried out. This presents a mystery, because by September 1938 the Westchester group had gone so far as to follow James Wyld's example by producing their own regeneratively cooled motor. Limber said that it embodied several innovations, the most notable being the introduction of "a liquid coolant within the combustion chamber."

In the April 1938 issue of *Astronautics,* he was more specific. He called the engine a "double chamber liquid-cooled motor." As it was also of 25 mm (one inch) diameter, it was considered "one of the largest to be tested on the new ARS proving stand." The stand was not used again until as late as the Summer of 1941 at Midvale, and at that time it was Wyld's improved regeneratively-cooled engine that received all the attention. The Westchester Rocket Society by this point, like the Yale and Peoria groups that preceded it, disappeared.[218]

In 1936 Capel W. McNash of Chicago, in company with B. D. Levi, went to the offices of *Popular Aviation* in the same city and approached the editor with their plan for the American Institute for Rocket Research. McNash was not new to rocketry. He had been a member, or spokesman, at least, of the Cleveland Rocket Society two years earlier when he was then an assistant manager of the United Press Bureau in Cleveland. McNash is also identified in a story about the CRS in the *Cleveland Press* for 8 December 1944 as then being "a lieutenant serving in the South Pacific."

In 1936 McNash and Levi told *Popular Aviation* that it was unfortunate, in view of the tremendous possibilities of rockets in the high speed transportation of mail and passengers, and for obtaining "samples of the stratosphere" for meteorological and cosmic ray research, that pitifully small backing had been provided in the interest of science in this country. The pair lamented the fact that the experimenters were forced to depend upon meager subscriptions and dues from societies and the sale of rocket stamps and postcards to collectors. McNash and Levi were determined to do something about it. *Popular Aviation's* editor could not agree more and subsequently published an outline of the AIRR's aims—"to attain their common goal of high speed transportation in the stratosphere by means of rocket motors."

McNash also tried to solicit help from Robert H. Goddard, but was characteristically rebuffed. No overwhelming response for the aims of the American Institute for Rocket Research was forthcoming from the readers of *Popular Aviation* either and hence, another would-be rocketry organization passed into oblivion.[219]

The Amateur Research Society faired no better. Founded in 1937 at Clifton, New Jersey, by Nicholas Swerduke, this organization of five or six engineering students set up its own laboratory, conducted a series of solid-fuel rocket experiments and began preliminary work with liquids. Like Limber's Westchester group, they too were situated close to the ARS and thus also became an affiliate. Headquarters were at 12 East Russell Street. Virtually nothing further was heard of this second "ARS", and it is assumed that they too succumbed to pecuniary strangulation and absence of real support.

Exactly the same could be said of another group which existed in 1937. The apparent reason for its anonymity was a self-imposed secrecy. It is known that about that time engineers of the nonprofit research organization, the Batelle Memorial Institute of Columbus, Ohio, in conjunction with Ohio State University (also in Columbus) conducted experimental rocketry but did not wish to taint their professional reputations. The 9 May 1937 issue of the *Cleveland Plain Dealer* published a story of the 26-year-old director of the "American Society of Rocket Engineers," Lester D. Woodford. Was this the secret Batelle-Ohio State group? The *Plain Dealer* identifies Woodford as an industrial engineer with the Addressograph-Multigraph Corp., and who had graduated from Ohio State University in June 1936; he had been experimenting with rockets since the age of 18, in 1929, and remained in the University for five years (1931—1936) because he wanted to take all the courses he could that related to his experiments. Woodford is also said to have *tried out his rockets in a quarry near Columbus.*

Popular Science for September 1932 has a picture article of Woodford in which it is vaguely suggested that his gasoline-liquid oxygen rocket was meant for aircraft propulsion. But the story also reports that "trying to develop a commercially practical craft, he [Woodford] also feels the perfected ship could fly to the Moon and back. The model carries an automatically operated parachute for safe landing." More interestingly, Woodford is said to be presently (1932) "on an island in a Canadian lake where the first tests with his model are being made." Woodford's collegiate records show that he was a professional man in every sense. He was a private aviator who was "carrying on research in the field of aeronautical reactionary heat engines," a former president of the Ohio State Aeronautical Society, assistant editor of the *Ohio State Engineer* and, later, Chief Industrial Engineer for the Ted Smith Aircraft Co. Yet the present whereabouts of this individual is unknown and an inquiry to Batelle has produced not one scrap of evidence of any rocket work in the 30s. The mysterious American Society of Rocket Engineers remains a secret.[220]

In the 1930—1940 period, three other American rocketry "organizations" came into being. One was Bernard Smith's California Rocket Society, started in the Summer of 1940; another was Robertson Youngquist's Massachusetts Institute of Technology Rocket Research Society, also begun in 1940. The third was a supposed rocket group formed at Tri-State College, Angola, Indiana. Smith's society, which afterwards conducted among the first hybrid propulsion experiments, falls outside the period under study. The same applies to the MIT Society, while the Tri-State College group never materialized.[221]

Finally, there was the Springfield Society for Rocket Experimenters. Virtually nothing is known of it other than this organization was founded at Springfield, Ohio, in 1939 and was re-organized in 1945. It is not mentioned in Cedric Giles list of rocket societies and only appears in "Some Capsule Notes on Rocket-Astronautical Societies" found in the Andrew G. Haley papers.[222]

Smaller Foreign Groups—Argentina, Great Britain, Holland, France, and Japan

Outside America four or five other nations similarly spawned their own smaller and lesser known rocket or interplanetary travel societies. Letters received by the VfR and the other larger astronautical organizations amply attest to the international scope of the astronautical movement during the 1920s and 30s. From South America the VfR received inquires from Montivedeo, Uruguay, and Buenos Aires, Argentina. Herbert Schaefer was one VfR member who knew Spanish fluently; he lived in Madrid as a child for eight years. It therefore fell upon him to be the translator, and in fact, correspondent. The Argentinian correspondent was Ezio Matarazzo, a first year chemistry student of the University of Buenos Aires who was well versed in some of Germany and Austria's key pioneers which he had translated for his own use. Yet Matarazzo's knowledge of astronautics and rocketry was still "not very profound." He hungered for still more information, especially details on the VfR's experiments. He had already started an astronautical magazine in 1932—the official organ of perhaps the first space travel group in his country and the first in Latin America. This organization was called, somewhat clumsily, the *Centro de Estudios Astronáuticos 'Volanzan'*, or *'Volanzan' Center of Astronautical Studies. Volanzan,* also the name of the journal, is a peculiar Spanish acronymn difficult to translate into English. It is something like "launched flight" (i.e., rocket-launched flight). Fellow Founders were the students Adelqui Santucci and Julio R. de Igorzabal.

Matarazzo continued his correspondence with Schaefer until as late as 1936, after Schaefer moved to America. But nothing further resulted. Matarazzo could report nothing new about the *Centro de Estudios Astronauticos,* and only spoke in generalities of a proposed series of two books on astronautics he hoped to write in collaboration with Schaefer. The books never materialized and the correspondence ceased. Nothing else is known of the astronautical movement in South America during the 1930s. Though interest in *viajes interplanetarios* (interplanetary voyages) is certainly evident by various articles and notices in *Aeronáutica Argentina* and other South American aviation periodicals of the mid-1930s.[223]

In Great Britain there were several smaller astronautical groups or would-be societies. In Paisley, Scotland, south of Glasgow, an enterprising 15-year-old schoolboy, John D. Stewart, initiated the so-called "Paisley Rocketeers' Society," or simply, Paisley Rocketeers, on 27 February 1935. With his fellow school chums the Paisley group had 20 members. Throughout the years Stewart—up to the present—launched literally hundreds of rockets, all in direct contravention of the prohibitive Explosives Law of 1875. For all this, the Paisley Rocketeers' contribution to either the space movement or even to the stratosphere rocket was without any impact whatever, because Stuart and others were strictly hobbyists or entrepreneurs selling rocket mail. More directly attuned to the space travel dream was the Leeds Rocket Society, supposedly started in the late 30s by H. Gatcliffe; though Erick Burgess insists that the founder was Harold Gottlieb who was encouraged by Burgess through correspondence with him. But the war effectively ended this group not long after it was formed—as it did to J. A. Clarke's Hastings Interplanetary Society (no relationship to Arthur C. Clarke). Sir James Jeans, the famous British astronomer and science popularizer, is also known to have initiated an "Astronautical Section" in his Junior Astronomical Association in 1938, but this too was not long-lasting.

A Combined British Astronautical Society also functioned and became the nucleus of the post-war BIS under the direction of Burgess and Gatland. It was an amalgamation of the Manchester Astronautical Association and the Astronautical Development Society. This latter group was founded in 1938 by Kenneth W. Gatland, later editor of the BIS magazine *Spaceflight,* and H. N. Pantlin. Both men were then employed as junior draftsmen in the drawing office of Hawker Aircraft Limited at Kingston-Upon-Thames. As such, the ADS was an outgrowth of the old Hawker Model Flying Club. The Society prospered somewhat and during the earliest phase of the war was able to build a test stand called a "Centrivelo." Small solid models were also flown— without arousing attention of the authorities. In 1944 the ADS joined with the MAA to become the Combined British Astronautical Societies, with Gatland as Secretary and Burgess as Chairman. War-time and post-war British developments have a more convoluted history, but it is enough to say that interest in astronautics had never been so strong in Britain as during this and the immediate post-war period. The CBAS had several hundred members and many local branches when it was disbanded and reformed into a post-war BIS. The British space travel movement sparked by P. E. Cleator had gathered momentum at last.[224]

Elsewhere in Europe, Dutch astronautical literature in the late 20s and early 30s was particularly good as evidenced by the private collection of materials from this period belonging to Maarten Houtman, a former editor of the Dutch Astronautical Society's *Spaceview* magazine. Perhaps this interest may be accounted for by the close proximity of that country to Germany and England, as well as of Dutch familiarity with both languages. Probably through the impact of this literature Gerard A. G. Thoolen began the *Nederlandse Rakettenbouw,* or *Stichting Nederlandsche Rakettenbouw* (Dutch Foundation for Rocket Building) on 17 October 1934 at s'Gravenhage (The Hague). According to contemporary Dutch newspapers, the NRB (as the Dutch abbreviate it), grew out of a *"Studie-Comité"* (Study Committee) that included "various aeronautic persons." A magazine was planned called the *Rakketenpost,* to be published by the firm of *Caspers' Uitgeversbedrijf* in the Hague. Shortly after the founding, Thoolen wrote to Pendray informing him about the club and adding that lack of finances then made it impossible to publish a journal. Some of the NRB's 20 members were nonetheless hard at work building a rocket.

Besides Thoolen, the original officers of the NRB consisted of Karl Robertti, S. Arkema, and P. Pastelein. It was soon discovered that contrary to their stratospheric and interplanetary ambitions, the NRB had quickly degenerated into a terrestrially-bound capitalistic venture, the selling of mail rocket covers. This was the most lucrative area of rocketry at the time. The German Gerhard Zucker, for example, was little more than a non-technical business entrepreneur with a slight touch of the visionary. One or two other mail rocket pioneers, particularly Friedrich Schmiedl of Austria, tended to be more technically oriented and more idealistic. Most were dubious pioneers. Thoolen and some of his associates unfortunately fit into this category. Very early, all pretense at studying interplanetary flight was cast aside. The respected Dutch philatelic journal, *Nederlandsch Maandblad voor Philatelie,* published a lengthy condemnation of Thoolen and his group entitling the article "Swindle With Rocket-Mail." On 6 December 1934, the journal reported the NRB's first "post-rocket" was fired on the shore near Katwijk aan Zee. This "new" rocket was nothing more than a modified skyrocket obtained from the fireworks manufacturer A. J. Kat of Leiden. It and all other NRB rockets added nothing to the advance of technology nor even afforded proper experience for the "experimenters" and so-called "inventors." In any event, the first rocket exploded. Whether the NRB's launches fizzled or not, all were treated as historical firsts with the colorful cachets always fetching handsome profits.

The grand result of Thoolen's activities was a jail sentence for several frauds, especially for the falsification of "rocket stamps." One of his co-workers, a Professor Dr. Adam J. de Bruijn— later found to have practicing under a false license—was similarly incarcerated. Stamps and covers sent in Karl Robertti's rockets were also highly suspect. There were legitimate if naive mail rocket pioneers to be sure. But generally this flurry of rocket activity had no real connection with the interplanetary travel movement per se, and in fact, detracted attention from it.

As such, rocket post does not fit within this study, though one more such group should be mentioned. This is the original Australian Rocket Society founded at Brisbane in 1936 by Alan H. Young and Noel S. Morrison. It too was a strictly postal rocket concern. Young made his first mail rocket flight on 4 December 1934 from the deck of the *S. S. Canonbar* to the city of Brisbane on the occasion of the arrival of the Duke of Gloucester. From then on this sort of showmanship fairly characterized the later flights of the Australians. Young's Society flourished until the beginning of World War II and should not be confused with the second Australian Rocket Society started in 1941 by J. A. Georges.[225]

Finally we come to the French and Japanese. The homeland of Jules Verne, Robert Esnault-Pelterie, and the REP-Hirsch Prize somehow never mustered enough support for its own astronautical society on a par with those in Germany, England, Russia, and America. Esnault-Pelterie himself lamented this when he spoke to Eric Burgess on Burgess' visit to REP's Esnault-Pelterie's Boulogne-sur-Seine home in 1937. According to Burgess, REP was "disillusioned" with his countrymen and remarked how lucky the Englishman was to enjoy comparatively ready support. The political difficulties in France during the 30s and especially the grave anxieties wrought by Hitler's open denunciation of the Versailles Treaty in 1935, account for some of this less than optimistic attitude in the country at the time.

Esnault-Pelterie, in any case, well knew of the efforts of Russian-born Alexandre Ananoff in attempting to start a group in France. The REP-Hirsch Prize Committee, begun late in 1928 and meeting for the last time in 1939, consisted of about 20 people whose sole purpose was to convene a few times a year to decide who was deserving of the honor of the annual REP-Hirsch award. This group could not be said to have constituted a true astronautical or rocket society. Ananoff thus independently sought to start a bonafide organization. He began his lifetime career in astronautics by authoring a treatise on "interplanetary navigation" in 1929. About 1930 he joined the prestigious *Société Astronomique de France*—the French Astronomical Society. From then on he was frequently mentioned in the Society's *Bulletin de la Société Astronomique de France.* His telescopic studies of lunar and solar eclipses were early noted, for example, and he wrote at least one article on his observations of Mars. Mostly, Ananoff was wholly wrapt up in space travel. In this he was significantly influenced by a fellow member and amateur astronomer Esnault-Pelterie. Soon, Ananoff was noted for his own talks upon the subject. Aided by 1926

Nobel Prize winner, physicist Jean Perrin, and probably also by Robert Lencement, he was successful in obtaining an "Astronautical Hall" (Hall 1010) in the Astronomy Division in the *Palais de Découverte* for the 1937 International Exposition in Paris.

The distinguished Secretary General of the *Société Astronomique,* Mme. Gabrielle Camille Flammarion, widow of the great astronomer and founder of the Society, had such confidence in the dynamic young man that she granted permission for him to start a *Section Astronautique* as part of the Society. The date of this founding, which Ananoff considered France's first astronautical club, was 9 May 1938. Mme. Flammarion permitted the group to meet regularly in the Society's library. Thus, the May 1938 issue of the *Bulletin* announced that: "For all work concerning astronautics, we request you to address Mr. A. Ananoff the second Monday of the months of June and July 1938 between 21 and 22 hours, in the Library Hall of the Society, rue Serpente, 28."

But perhaps Ananoff derived some of his inspiration, if not a concrete example, for his group from the Germans. In his *L'Astronautique* (1950), Ananoff recalled that "Until 1934 the closest collaboration existed between ourselves and the German astronauts [sic], then correspondence grew thin to abruptly stop." This, of course, amounted to letters between himself and the old VfR that flourished until 1934. Even after this, Ananoff received considerable encouragement and guidance from Hans K. Kaiser of the *Gesellschaft für Weltraumforschung.* In the case of the GfW the cue was more direct. Kaiser's organization set the pattern by growing out of a larger *astronomical* society, this development occurring a year before Ananoff started his own group. Kaiser's journal *Weltraum* gave unmistakable proof of a close cooperation, age-old enmities between the two nations being quite ignored. Kaiser wrote in *Weltraum* for January 1939 that: "As Mr. Ananoff reports, in France there reigns so great an interest in the work of the *Section Astronautique* as was not observed during the prime of rocket researches ten years ago. Mr. Ananoff is to be thanked again here for his interesting reports. May the young French Society achieve success in the future!" Did the dream of space flight truly transcend nationalistic barriers? Or was Kaiser acting out of self-interest in welcoming an expansion of the astronautical movement so that the chances for attaining the ultimate goal would be that much closer?

Certainly the promise of a mutual exchange of ideas and keeping the dream alive were also positive and at the same time idealistic steps. What transpired in the mind of Kaiser, the German, from late 1939 until 1942 when he found himself working on the V-2 rocket project directed against the countries he had so diligently tried to cultivate is a provocative question. The answer seems to be that as in the case of Ananoff, the Frenchman, the dreams indeed transcended national barriers and politics: both men were very active in the first International Astronautical Federation congresses after the war.[226]

Of the specific activities of Alexandre Ananoff's *Section Astronautique* we have the benefit of a contemporary letter he wrote to Willy Ley on 20 October 1938 in which, after describing other activities, he says: "Sometime in March or April of 1939, I plan to launch a rocket in the vicinity of Paris, at Saint Cyr, on a two-kilometer strip, in trying to make launching more stabilized and precise. This experiment will also be a real spectacle for the general public, in order to give some importance to the Astronautic Section."[227]

The *Section Astronautique* showed promise and had gotten off to a good start, but before very long it expired. This was neither the fault of Ananoff nor the spectre of war upon the horizon. It was an internal matter that Ananoff is at a loss to completely explain. He only reports with ill disguised acrimony that certain unnamed persons within the *Société Astronomique de France* harbored resentments and prohibited the proper conduct of the Astronautical Section's business. In his *Les Mémoires d'un Astronaute ou l'Astronautique Française* (1978) (Memoires of An Astronaut or French Astronautics), he reveals that soon after the Section was inaugurated, the meetings became "very irregular" and that the "pretexts were multiple and varied." At one time, all the chairs in the meeting place were removed without explanation! At another time—the last attempted meeting of the group—the door was locked and the key could not be found.

This was the end of the *Section Astronautique,* but it was far from the close of Ananoff's own involvement in astronautics. He continued to produce a steady stream of astronautical articles and to correspond with others in the field even though the following year saw his nation swiftly conquered by Germany; Ananoff himself briefly served as a telephonist with an artillery regiment. He attempted to interest the *Ministère de l'Armement* in rockets as defensive means, but without success. Then, he found himself a captive of the Germans, being interred in *Stalag*

XIII D. The national and personal hardships still did not deter him and he continued to write. Interest in interplanetary travel also remained alive in the minds of other Frenchmen despite the war. The *Bulletin de la Société Astronomique de France,* published at least two items on the subject during this period.

At the war's end Ananoff again devoted almost all of his waking hours in the pursuit of the promotion of astronautics. In 1950 his organizational abilities saw him succeed far more than he or his colleagues had dared realize. He recalled that on 19 November 1937 André-Louis Hirsch had written to him that: "To organize an international astronautical congress is an impossible task to realize and in any case the obstacles are innumerable, I assure you . . ." In 1950, almost single-handedly, Ananoff created the International Astronautical Federation whose first congress was held that year in Paris. But this is a larger story which is now fully covered by Ananoff in his memoires.[228]

By contrast, the Japanese astronautical picture in the 1930s is virtually unknown, or a puzzle at best. The June 1935 issue of *Astronautics* published a curious photograph from Acme Newspictures, Inc. showing, amidst a huge Tokyo crowd, a very large rocket said to have been designed by Mr. Tsunendo Obara and exhibited the previous October. The original news photo further reveals that the rocket, antennas and all, was on exhibit on or about 6 October 1934; that the Japanese characters around a sort of porthole on the top read "Rocket, modern style, Air Force", and that Mr. Obara was from "The Nippon Rocket Society." In his survey of rocketry activities around the world published in 1937 in the *Bulletin de la Société Astronomique de France,* Ananoff mentions only Obara's project under the heading of Japan and also adds that: "Information relative to these experiments is lacking." The English-language *Japan Times & Mail* for 6 October 1934 does contain a rocket article, but dealing only in a generalized way with "rocket artillery" as the "latest menace to civilization" and not with any Japanese experiments. Part of the mystery of Obara's project is cleared up from a letter sent to the ARS' journal *Astronautics* by Tatsue Hasegawa of the *Nippon Sanso Kabushiki Kaisha* (Japan Oxygen Co., Ltd.) of Tokyo. The letter, dated 17 July 1935 says: "Regarding that experiment [which]appeared in newspapers several months ago, I should say it does [sic] not worth mentioning, so far as I could learn at that time, the model was of such a primitive construction, using gun powder for the fuel. It was reported that the model failed to fly at all. I was not interested even to attend the experiment." Nothing else is known of Japanese astronautical activities.[229]

We conclude with another youthful group. Kurt Stehling, a refugee from Hitler's Germany, formed at age 15, a rocket club at the Central Technical School in Toronto, Canada, in 1936. Much like their British counterparts in Eric Burgess' Manchester Interplanetary Society, the Canadian rocketeers, numbering about 20, constructed small gunpowder-propelled rocket models. Fortunately, Stehling's club remained unhindered by any antique but enforceable Explosives Act. Besides flying rockets, the club assembled cardboard space ship models and corresponded with the BIS, ARS, and other groups and individuals. Stehling's activities were impressive enough to merit an interview in 1939 on Toronto radio station CFRB. But the war put a halt to his rocketry as it did with the larger rocket and astronautical organizations. This was only a temporary setback. Like Burgess, Stehling had merely completed his "apprenticeship" in rocketry and astronautics and was fully prepared to make these subjects his lifelong pursuits. Almost immediately after the war he established an organization that became the Canadian Rocket Society. Later, he became a leading member of several international societies.[230]

In reaching for their often fanciful goals of designing and constructing stratospheric rockets and spaceships, these large and small groups of the 1920s and 30s did not always take the right paths. But they always caught the attention of the public. In the process, they established a certain public acceptance and infectious enthusiasm for the possibility of spaceflight where none had existed before. In retrospect, this was their greatest achievement.

IX

The Contributions of the Rocket Societies

A concluding fundamental question must be: what was the actual impact of these groups? Sociologist William Bainbridge provides one surprising answer in his *The Spaceflight Revolution, A Sociological Study* (1976). Clinically and eloquently, he analyzes two of the early societies, the AIS/ARS and BIS, and the figure of Wernher von Braun. The penetrating sociologist's eye of Bainbridge makes the fascinating observation that the phenomena of the early societies was nothing short of a "revolution." Bainbridge's definition of revolution was sociological and political. Spaceflight, he says, was achieved by a handful of people who sold governments the necessity for it, "despite the world's indifference and without compelling economic, military, or scientific reasons for its accomplishment. Not the public will, but private fanaticism drove men to the Moon." The process began in the 1930's.

One of the pivotal proofs of Bainbridge's argument is the role of Wernher von Braun and his Peenemünde team. The first stage of the revolution, or successive revolutions as Bainbridge sees it, was von Braun and others of the old VfR convincing the German Army of the need for a super-artillery—a large liquid-fuel bombardment rocket—though it was not cost-effectiveness vis-à-vis conventional weapons. All the while Bainbridge contends, von Braun and his VfR associates at Kummersdorf and Peenemünde had their heads turned towards space. In short, von Braun and his people manipulated the German war machine into paying for the development of their interplanetary spaceship. Bainbridge sees the subsequent revolutions in a similar light: small numbers of star-struck men (in which von Braun in the US, and Korolev in the USSR were prominent), continuing to sell the government or a disinterested public on the necessity of harnessing the new super rocket hardware for the astronomically expensive gamble of the conquest of space.

Bainbridge's conclusions are most interesting but seem only partly valid. A closer examination tells a different story. Bainbridge's prima facie argument is somewhat diluted when we recall that the German Army was already interested in liquid-fuel rockets prior to approaching von Braun. Bainbridge does mention this and even stresses that the Army (in the person of Professor, Colonel Karl Becker, head of the Army Ordnance Office) "could not have turned to liquid-fuel engines had not the Spaceflight Movement prepared the way." This should have been considered the real beginning of the revolutionary chain. Becker was, after all, a key man in the German Army who was already "sold" on the liquid-propellant rocket some years prior to 1930—before von Braun joined the German Rocket Society.[231]

It was fortuitous that von Braun arrived on the scene when he did and that he possessed serious technical devotion to rocketry coupled with organizational skills, level-headedness and diplomacy. As it was, Colonel Becker and Captains Dornberger and von Hörstig had to wait for a good two years before almost giving up on their own efforts to develop an efficient liquid-propellant rocket engine before turning to the VfR for possible technical help. Their *discovery* of the talented von Braun was another very important element in the turning of the course of events towards the development of the space rocket.

It will be remembered that when the Army approached the VfR, Rudolf Nebel, a supreme but deceitful opportunist with little actual technical ability, came forward with his own terms in which the Army was to pay him to undertake experiments and to support the publicly-known *Raketenflugplatz*. The young von Braun at once saw the irony that those who did not have the money should hardly dictate the terms. He had a heated discussion with Nebel about this point and was later proven correct. Nebel's obvious showmanship, technical ineptitude, and private deals were not what the Army wanted. The officers made it clear that they were not interested in furthering spaceflight. Besides, Nebel's test rocket hardly worked. On the other hand, von Braun's tact and engineering abilities did impress the military and when they offered to hire him on their terms of absolute secrecy he accepted. Von Braun was a pragmatic individual acting positively upon an offer made to him in his interests so that he could still pursue his rocketry, though in secrecy and for military purposes.

It is incorrect to suggest that there was some sort of collusion on the part of von Braun and other VfR members to misleadingly sell the military on liquid-fuel rockets. There simply was none. Other VfR members later did join von Braun to work for the Army, but comparable to von Braun's case, they could actually get paid for pursuing their avocation of rocketry. It was a golden opportunity in the face of the Depression.

The point was made at the beginning of the book that not all who joined the VfR or other societies were necessarily space-minded. Some were simply fascinated with the engineering potentials of the rocket. Nor was there a master plan for reaching space, much less one for manipulating the Army to develop a big rocket that could be converted into a spaceship in the future. Von Braun and his former VfR associates really did have no idea where their work would lead. Von Braun was genuinely "surprised" at the turn of events in their favor. And if von Braun and his military bosses sometimes resorted to questionable tactics to get more and more money, men and materials for their rocket project, the cause of spaceflight was not necessarily behind it. Becker and Dornberger took a greal deal of pride in their work just as much as von Braun. All were jealously dedicated to see it through and this often meant overcoming military and bureaucratic hurdles as best they could. These were also times of fierce competition and other priorities in the German war machine. Sometimes, however, the *Wehrmacht* heirarchy itself was the prodder. Dornberger remembered that "as we kept on pestering the Army chiefs for money for continued development we were told we should get it only for rockets that would be capable of throwing big loads over long ranges with a good prospect of hitting the target."[232]

Yet it cannot be denied that von Braun and some of his former VfR colleagues did harbor the dream of spaceflight. But in a real sense this dream had to be apart from the military work at hand. This was especially true when the war broke out.

It is interesting nonetheless to briefly explore von Braun's early spaceflight visions as they are not only germain to the subsequent development of the spaceflight movement of the 1930's but also because von Braun's outlook paralleled those of the Russian Sergei Korolev and the American Robert Goddard.

In 1931 when he was 18, von Braun wrote a school paper titled *"Lunetta."* It was a plotless science fiction account of what a space station might be like in the future. The paper was obviously based upon Hermann Noordung's *Das Problem der Befahrung des Weltraumfahrt* (The Problem of Space Travel Flying), published in 1929. By his own admission von Braun remarked that during those years he "read everything in the space field, including Willy Ley's popularizations." This seems proof enough he was won over early to the spaceflight dream. In the Spring of 1930 he joined the VfR and quickly rose to prominence in the organization. Two years later his first published writing on the subject, *"Das Geheimnis der Flüssigkeitsrakete"* (Mysteries of the Liquid-Fuel Rocket), appeared in the popular science magazine *Umschau* (Review) for 4 June 1932, a few months before he entered the payroll of the German Army. A scant two lines are devoted to space travel. The rest centered on the workings of the rocket.[233]

The dream of spaceflight had not withered in von Braun's mind. But by 1932 he already had considerable experience actually building and firing rockets and their technological perfection most preoccupied him. The spaceflight idea became subordinated to the perfection of the rocket, particularly in the light of the disappointingly feeble power the liquid-rocket was able to produce in the early 30's.

Robert Goddard went through precisely the same process of de-emphasization of the spaceflight idea the more he labored to develop the basic liquid-fuel rocket hardware. The American Interplanetary Society also passed through this transformation: for this reason as well as others they became the American Rocket Society. In all cases there was no abandonment of the spaceflight dream but only a maturing realization of the enormity of the technical problems that lay ahead. In the USSR, Sergei Korolev also instinctively lowered his sights. We saw in the Russian chapter that he cautioned the writer Yakov Perelman to downplay the yet unreachable goals as going to the Moon and planets, and to focus his writings mainly on rockets and the possibility of stratospheric rocket planes. Korolev complained about how difficult it was to put forth his own rocket ideas while at the same time maintaining respectability in the scientific community. Here was the crux of the matter which was common to all the rocketeers.

Korolev, Goddard, and von Braun stood above the rest because at the beginnings of the careers they quickly learned to stay aloof publicly from the mainstream of the early spaceflight movement, at least in its interplanetary flight aspects. This was the only way they could succeed—and all of them did—in getting the kind of real support they needed to undertake their rocketry experiments. In this respect, Bainbridge's theory is correct, yet at the same time it was just as true that spaceflight ceased to be in the forefront of their thinking.

Von Braun's early association with the VfR was public knowledge in Germany and there is the well-known story that he, along with fellow Peenemünde staff members Klaus Riedel and Helmuth Gröttrup were arrested in 1944 for actually subverting the V-2 program because they were really working towards the development of the spaceship. Yet even Bainbridge admits that the case was not clear because von Braun had refused to cooperate with Secret Police chief Heinrich Himmler when Himmler tried to take over control of Peenemünde. Generally, there seems no evidence that von Braun openly espoused spaceflight while working for the German Army. Objectively speaking, the answer to the question of von Braun's long-range goals when he entered German Army employment remains unanswered, but the fact that the V-2 led to future space vehicles is undeniable.

Bainbridge, in his *The Spaceflight Revolution,* also very briefly covers the role of Sergei P. Korolev in the USSR's rocketry and spaceflight history. He dwells mainly on Korolev's possible influence with the Soviet Government in undertaking its space program. Without the benefit of sufficient Soviet documentation, however, Bainbridge misses the important point that after the early Russian rocket organizations (MosGIRD and LenGIRD) experienced a brief civilian status, the Soviet Army's top command—in the person of Marshal Tukachevsky—directed and financed rocket programs for its own ends. Very early the spaceship idea was subordinated to weaponry or other military applications. That situation persisted for many years thereafter.

The eventual steps that led both the US and the USSR to enter space exploration in the post-war era are more complex issues that do not fall within the scope of this book. Perhaps Bainbridge's "revolutionary" hypothesis is more valid here though there are several other books that also address themselves to these questions, examples of which are John M. Logsdon's *The Decision to Go to the Moon* (1970) and Nicholas Daniloff's *The Kremlin and the Cosmos* (1972).

Bainbridge's term "revolutionary" is more aptly applied to the work of the early space travel societes themselves. It was an intellectual or sociological revolution: they made the world space conscious. The societies used every modern means of communication to proselytize their work and aims—public talks, newspapers, magazines, radio, newsreels, exhibitions and even nascent television. It almost did not matter how wildly exaggerated some of the Sunday supplement and popular stories were of the societies and the space travel movement in general; such distorted stories may have helped arouse public interest that much more. To paraphrase former Smithsonian Fellow Dr. Joseph J. Corn, in his thesis "A New Sign in the Heavens: The Gospel of Aviation and American Society, 1880— 1950," the early rocketry and space travel societies "made space travel thinkable to the masses." We may also cite other similarities in Corn's thesis—that astronautics, like aeronautics during the 1920s and 30s, was similarly promoted by its "believers" to almost the pitch of a religious crusade; both had their "evangelists," proclaiming the airplane "for the common man" as just around the corner or rocket travel from coast-to-coast and even to the planets as inevitable and imminent.[234]

It is also "revolutionary" that the gradual, albeit never completely fulfilled molding of public opinion to accept rockets and the possibility of space travel, was undertaken during the darkest days of the Depression. Here, we may recall the pithy remark of the early ARS member Bernard Smith who admitted that he joined in the 1930s because the world was "a lousy planet and I wanted a way to get off it." Interplanetary travel was as good a means as any of escape.[235]

On the other side of the coin, the early rocket and space travel pioneers at first faced skepticism and even ridicule by the general public and the scientific community. It was revolutionary that the space travel societies changed these attitudes at all; the battle for public and scientific acceptance was never fully won, but through the years inroads were made. P. E. Cleator and others regularly lashed out at the press for their sometimes ill-disguised doubts as to the sanity of "the Moon men" and "star-struck" rocketeers. Looking back from the vantage point of Apollo Moon-landing days, surviving early space travel society members take no small pride in recognizing how farsighted they really were, especially in the face of criticism and ignorance. They are also proud that they were indeed "odd" at the time, as are all prophets of science. Val Cleaver, the late and respected British rocketeer, described the pre-war BIS as "a small group of enthusiasts and cranks." "And why not?" asked H. E. Ross, another early BIS regular. "We were unorthodox." In Germany *Raketenflugplatz* founder Rudolf Nebel defiantly entitled his book *Die Narren von Tegel* (The Fools from Tegel), after a reporter's pointed characterization of the men of the *Raketenflugplatz.* The Soviet side is harder to pin down as public opinion is never regarded;

the histories of Soviet space and rocketry achievements dating back to the GIRD and pre-GIRD days are also deliberately glorified. But undoubtedly Soviets too struggled against more conservative viewpoints and had to contend with governmental dictates and purges besides.[236]

One example of early attitudes on space travel is found in a letter from American Interplanetary Society member Noel Deisch of Washington, D.C., to the Society's President, David Lasser, dated 1 March 1931: "Although there undoubtedly must be a number of scientists and technical men in Washington who are interested in astronautics, I really am not acquainted with a single person of the calibre you require, who has a serious interest. Recently, on a visit to the Carnegie Institute here, I took the liberty to ask a prominent visiting astronomer from [the] Mt. Wilson [Observatory, California], who is doing research on the Moon, whether he thought Dr. Goddard would succeed in getting there and making observations on the site itself. He was completely taken back at the question, and could hardly find words to express his amazement. After he had recovered he stated that he was positively sure, and on purely ballistic grounds, that nobody would ever get there!"[237]

Despite difficult times, early space travel and rocket societies contributed not only a change of opinion towards spaceflight and gave new visions of hope, but also provided inspiration and motivation. All the major groups received innumerable inquiries on how to join, how one could help the cause and how to build rockets and start their own societies or chapters. The societies were also learning, or more accurately, exchange centers as the field was so new that everybody could learn from each other.

The Soviets call Korolev's GIRD days his "apprenticeship." Many young, future rocketeers turned to these groups for both guidance and experience. Helmut M. Zoike, later a top flight engineer in charge of development and testing the V-2 power plant at Peenemünde and afterwards one of von Braun's Saturn team, was initially inspired by reading Hermann Oberth's 1929 classic *Wege zur Raumschiffahrt* (Ways to Space Travel). He then joined the VfR's *Raketenflugplatz* in October 1930 at the age of 15. Dr. Weingraber, a family friend who managed a machine department at Siemens & Halske informed young Zoike about the *Raketenflugplatz* which immediately excited the boy. ". . . I could now see a way on [sic] how I could practically contribute to this great dream of mine," Zoike recalls.[238]

Many examples of individuals who stayed with rocketry and space travel to contribute to it technologically, and other ways, are found in this book. They too received their "apprenticeships." By name they are: Wernher von Braun, Serqei P. Korolev, Hermann Oberth, Johannes Winkler, Val Cleaver, Alfred Africano, Klaus Riedel, Rolf Engel, Hans Hüter, Kurt Heinisch, Heinrich Grünow, Herbert Schaefer, Hans K. Kaiser, Albert Püllenberg, Alexandre Ananoff, Willy Ley, Arthur C. Clarke, Kurt Stehling, Edward H. Seymour, John Shesta, H. Franklin Pierce, James H. Wyld, Lovell Lawrence, Leonid S. Dushkin, Valentin P. Glushko, L. K. Korneyev, Igor A. Merkulov, Yuri A. Pobedonstev, Mikhail K. Tikhonravov, Frank J. Malina, Krafft Ehricke, Robert C. Truax, Eugen Sänger, Robertson Youngquist, Guido von Pirquet, Franz von Hoefft, H. E. Ross, and Ralph A. Smith. This does not purport to be an all-inclusive "who's who" of rocketry and space travel of the 1920s and 30s, but does demonstrate the influence of the societies and how they provided a forum and "schooling" for these men. Through the societies, and their exchanges of literature, correspondence, speakers and visits, the influence of these pioneers and the spread of their ideas were greatly magnified.

These ideas sometimes had far-reaching concepts which opened up possibilities for future areas of space research. The BIS journal's early reports of Karl G. Jansky's discovery of radio disturbances from outer space did much to educate, stimulate interest and perhaps build support for the later field of radio astronomy. Direct technological contributions of the societies, as we have suggested, are more modest and even elusive. The training afforded some of the pioneers was far more invaluable than actual hardware produced. There was also a considerable literature built up through the publication of society journals and reports so that even negative results· might prove "useful." One direct spinoff, however, was the formation of Reaction Motors, Inc., which became one of the largest and important American aerospace companies. It was started by four members of the ARS—Wyld, Shesta, Pierce, and Lawrence—and based upon Wyld's significant advance of the regeneratively-cooled motor that was repeatedly tested and refined on the ARS test stand No. 2.

Soviet technological gains from 1930s activities are far more difficult to trace. But certainly many Soviet and world wide "firsts" can be traced to the early groups: the GDL electric rocket engine, GDL's early JATO's, gasoline and nitric acid, and other early "exotic" propellant combinations (as in the ORM-12 motor of 1932), and so on. In the USSR there was enough of the 1930s school of rocketeers legacy that the first *Sputnik* vehicles may well have derived from native roots rather than captured V-2's.

Even if the technological "spin-offs" of the early societies are sometimes uncertain, there is no question these highly dedicated and persistent organizations sustained the dream of spaceflight. The war interrupted their efforts only temporarily.

Upon cessation of hostilities not only were the BIS and ARS born anew and flowered into greater organizations, but new societies sprang up everywhere. This renaissance was in no greater evidence than in the earliest International Astronautical Federation congresses, starting with the first in Paris in 1950. Total attendance was then over 1,000, with delegates from many European countries; one of the overseas delegates was Ing. Teófilo M. Tabanera of Argentina who himself had written upon space travel as early as 1932. From 1950 on, the growth of participating members and societies to IAF congresses has been phenomenal, truly vindicating the need and viability of these groups from the beginning.

The late Andrew G. Haley, writing in 1958, and Frederick C. Durant, III, writing in 1950—61, compiled directories of these organizations of that period. Even at that relatively early time the statistics are impressive. In 1950 the ARS reported more than 17,000 in their membership, with regional sections from coast-to-coast. Their journal, *Jet Propulsion,* was considered the finest technical magazine of its kind. The BIS reported about 3,500 members world-wide; *Deutsche Gesellschaft für Raketentechnik und Raumfahrt* (the German Society for Rocket Engineering and Space Travel) had 1,600 members; the French Astronautical Society, 580 members; the Argentine Interplanetary Society (formed in 1949 by Teófilo Tabanera), 500 members; *Österreichische Gesellschaft für Weltraumforschung* (the Austrian Society for Space Flight Research), 144 members; with similar groups in 27 other countries. Obviously the launch of *Sputnik 1* on 4 October 1957 changed the whole picture of astronautics overnight. Intense world interest saw the immediate creation of several new astronautical societies (i.e., from 1958). Yet precedents for astronautical and rocketry groups had been set long before—in the late 1920s and 1930s, and saw a steady growth after the war.[239]

One of the most important accomplishments of the early groups were efforts towards international cooperation. Mutual exchanges of journals, visits, Werner Brügel's attempted IRK (*International Rakete Karterei,* or International Rocket Travel Information Bureau), and the International REP-Hirsch astronautical prize attest to this. The yearning for the international solving of the dream shared by all the societies is no more simply expressed than in Hermann Oberth's letter of introduction to Robert H. Goddard, dated 3 May 1922: "Already many years I work at the problem to pass over the atmosphere of our earth by means of a rocket. When I was now publishing the result of my examinations and calculations, I learned by the newspaper, that I am not alone in my inquiries and that you, dear Sir, have already done much important work in this sphere. . . . I think that only by common work of the scholars of all nations can be solved this great problem."[240]

This book has taken us through the first steps towards the fruition of that "great problem"—the interplanetary movement of the 1920s and 30s and the formative years of the first societies collected to achieve this goal.

It was the beginning.

Appendix

Table A. VfR (German Rocket Society) Rocket Experiments, 1930—1931.

Table B. American Interplanetary Society (American Rocket Society) Flight Rocket Experiments, 1932—1934.

Table C. Soviet Sounding or Scientific Research Rockets Actually Launched, 1933—1939.

Table A.

VfR Rocket Experiments, 1930—1931

Rocket Motor	Date	Performance	Remarks
Mirak I ("Minimum Rocket")	June 1930	Weighed up to 3 kgs (6.6 lbs.) (several models) Up to 4.5 kgs (9.9 lbs) 30—60 sec. thrust; held 1 liter (1 qu.) of fuel	Made by Riedel, Nebel, and Heinisch, Bernstadt Saxony, before *Raketenflugplatz* founding; long, thin aluminum tubing for tanks, cast aluminum head, chamber like the *Kegelduse;* carbon dioxide-fed; cooling by placing combustion chamber in oxygen tank, but unsatisfactory; no safety valves; explosions.
Mirak II	About April 1931	Higher weight and performance, not disclosed.	Escape valve installed for oxygen pressure; design of motor changed but neither Mirak I or II motors proved to be properly designed; explosion; not ceramic-lined as incorrectly reported by Pendray.
Mirak III (called by Ley "the egg" motor because of its similarity in shape to the prehistoric Aepyornis bird egg discovered in Madagascar in 1930. Final form also called 160/32 motor	About April 1931	This rocket was capable of 4.8 km (3 mi) altitude but not flown as no provision for parachute; weight, total rocket, 4 kgs (8.8 lbs), motor 85 grams (3 oz.); final form consumed 160 grams (5.6 oz.) fuel per sec. for 32 kgs/70 lbs thrust for about 30 sec.	
Repulsor I ("Two stick Repulsors" because of fuel nitrogen tank placement)	10, 14 May 1931 & other flights	18.3 m (60 ft) and 45.7—61 m (150—200 ft) flights; other figures not disclosed	With parachute; first VfR rockets to fly; water-cooled motor; parachute failures; designed by Klaus Riedel; term "Repulsor" chosen by Ley from the space rockets in Kurd Lasswitz' science fiction novel, *Auf Zwei Planeten* (On Two Planets), 1897.
Repulsor II	23 May 1931	61 m (200 ft.) flight; range, 600 m (1,970 ft.)	Oxygen fed by its own gas pressure and gasoline fed by nitrogen; parachute pyrotechnical release mechanism functioned; considered successful flight; designed by Klaus Riedel.
Repulsor III	June 1931	640 m (2,100 ft.) flight and three other good ascents with other models.	Greatly improved, though still parachute difficulties.
Repulsor IV ("One stick Repulsors")	Aug 1931 with tests on this configuration lasting until at least a year	1 km (3,300 ft.) to 1.6 km (1 mi); in one wild flight one went 4.8 km (3 miles) down range; one of the advanced one-stick Repulsors flown Aug. 1932, had a diam. of 10.2 cm (4 in), take-off weight of 20.4 kg (45 lbs), a thrust of 60 kg (130 lb)	Consciously used the Congreve method of placing the guidestick (tanks) in center line of rocket; water cooling jacket; parachutes functioned well; several flights; later Repulsors built by Riedel, et. al. used alcohol/liquid oxygen combinations, first 40% then 60% alcohol; also used regenerative cooling on some models, late 1931; exhaust velocity averaged 1,700 m/sec (5,600 ft/sec).

Table B.

American Interplanetary Society (American Rocket Society) Flight Rocket Experiments, 1932—34.

Rocket	Designer(s)	Date Flown or Tested	Location	Performance	Dimensions
ARS No. 1 (Technically should be called AIS No. 1)*	G. Edward Pendray and Hugh F. Pierce patterned after German Two-Stick Repulsor.	12 November 1932	Farm, near Stockton New Jersey.	Not flown; burned satisfactorily 20—30 sec. per 27 kg (60 lb)	Overall 2 m (7 ft;) Weight 6.8 kg (15 lb.)
ARS No. 2 (Technically should be AIS No. 2)	Bernard Smith; made from tanks and motor of ARS No. 1	14 May 1933	Great Kills, Staten Island, New York.	76 m (250 ft.) altitude per 2 sec.; thrust, about 27 kg (60 lb)	1.8 m (6 ft;) weight loaded 6.8-8 kg (15—18 lb.)
ARS No. 3 (Technically should be ARS No. 1)	G. Edward Pendray and Bernard Smith	9 September 1934	Marine Park, Staten Island, N.Y.	Not flown	1.2 m (4 ft;) overall diam. 20 cm (8 in) diam. long tanks 6.5 cm (6.5 in.)
ARS No. 4 (Technically should be ARS No. 2)	John Shesta, Laurence Manning Carl Ahrens and Alfred Best; made by Shesta	10 June 1934 9 September 1934	Marine Park, Staten Island, N.Y.	Non-flight test; Flight: landed 407 m (1,338 ft) 304 m/s (1,000 ft/s.)	2.3 m (90 in.) tall; motor about 12.7 cm (5 in.) diam.
ARS No. 5 (Technically should be ARS No. 3)	Hugh F. Pierce, Nathan Carver and Nathan Schachner			Project Abandoned	

*Based on name change of Society to American Rocket Society on 6 April 1934.

Table C
Soviet Sounding or Scientific Research Rockets Actually Launched, 1933—1939

Rocket	Designer(s)	Date	Dimensions	Results	Remarks
GIRD-09	GIRD team, principal designer M. K. Tikhonravov	17 Aug 1933	2.4 m (7.8 ft) long; takeoff weight, 19 kg (41.8 lbs). 180 mm (7 in.) diam.	400 m (1,312 ft) 1,370 m (4,500 ft), later model	First hybrid rocket, USSR; lox/solidified gasoline motor, 52 kg (114.6 lbs) thrust/15— 18 sec.; nine GIRD-09's built and flown, marking practical development of liquid rockets in USSR
GIRD-X (also called GIRD-Kh)	Initial design by F. A. Tsander	25 Nov 1933	2,200 mm (7.2 ft) high; 180 mm (7 in.) diam.; 29.5 kg (65 lbs) take-off weight	75—80 m. (246—262 ft.)	Lox and ethyl alcohol, first true Soviet liquid rocket; 70 kgs (154 lbs) thrust/22 sec. (one reference says 12—13 sec); basis of development of advanced Soviet rockets, 1935—1937.
M. V. Gazhala High Altitude Rockets	Initially designed by V. V. Razumov of LenGIRD for Leningrad Geographic Institute	1932	2 m (6.9 ft) long; 0.23 m (0.75 ft.) dia.; loaded weight 30 kgs (66 lbs), 5 kg (11 lbs) payload	Up to 1 km (3,280 ft)	Scaled down version of V. V. Razumov's 10 km (6.2 mile) rocket; smokeless pyroxylin fuel; engine designed by V. A. Artem'yev; how many ''high altitude'' versions unknown as leaflet-carrying and shrapnel simulation types also flown.
Aerodynamic-test vehicle for Razumov-Shtern Recording Rocket	Solid engine by V. A. Artem'yev; rocket body designed by V. V. Razumov	1934	2.6 m (8.5 ft) long; 0.35 m (1.1 ft) diam. weight of solid-fuel version unknown; liquid (unbuilt) version 90 kg (198.4 lbs)	No details given	Liquid version with complicated rotary engine designed by A. N. Shtern; liquid engine never fully built, project abandoned; planned payload of 31.4 kgs (69.2 lbs).
V. S. Zuyev Stratospheric Rocket	V. S. Zuyev RNII	1933—1934	Unknown but designed for 50 km (31 miles)	Original rocket not flown but design with 02 engine made flights	Possibly fitted with RNII 02 alcohol/oxygen engine of 100 kgs (220.4 lbs) thrust; this engine used on 216 winged rocket
Osoaviakhim R-1; later models called R-06 and ANIR-5	A. I. Polyarny	1934—1939	1,700 mm (5.5 ft long), ANIR-5 1,285 mm (4.2 ft); 126 mm (4.9 in.) diam.; loaded weight 10 kgs (22 lbs); payload 0.5 kgs (1 lb)	Unknown except that R-06 version had horizontal range of 5 km (3 miles)	Original rocket for Osoaviakhim organization was for meterological studies; when transferred to KB-7 (Design Office 7) was designated R-06 and may have been prototype for contemplated bombardment rocket; ANIR-5 version, shorter length and contained gyroscope for stability.
R-05 Sounding Rocket	A. I. Polyarny and P. I. Ivanov	1937—1939	2,250 mm (7.3 ft) long; approx. 200 mm (7 ft) high; ''two-stage'', total weight 60.5 kgs (133.3 lbs); 2nd or sustainer stage, 55 kgs (121.2 lbs)	Scaled-down liquid models flown, summer 1938, results unpublished	Liquid M-29e engine by F. L. Yakitis; Full version, never flown, was designed for 50 km (31 mi) KB-7 (Design Office 7) main responsible organization; elaborate payload, stability and guidance mechanisms worked out by other organizations; R-05g design version may have been a military vehicle.
Aviavnito sounding rocket	M. K. Tikhonravov and others	1936—1937	3,155 mm (10.3 ft) high; 300 mm (11.8 in.) diam.; weight 97 lbs (213.8 lbs); payload 8 kgs (17.6 lbs)	About 3,000 meters (9,843 ft), highest known flight, 15 Aug 1937	12-K ceramic-lined alcohol/lox engine designed by L. S. Dushkin, 300 kgs (660 lbs)/60 sec.

References

A. Interviews

Ackermann, Forrest J.
  1976.   By Frank H. Winter, 28 December 1976, Los Angeles, notes in author's collection.
Africano, Alfred
  1976.   By Frank H. Winter, 31 December 1976, Redondo Beach, California, tape in the National Air and Space Museum.
Beattie, John A.
  1979.   By Frank H. Winter, 9 February 1979, telephone, notes in "U.S. Rocketry, 1930's, General" file, National Air and Space Museum.
Burgess, Eric
  1977.   By Frank H. Winter, 20 October 1977, Washington, D.C., notes in author's collection.
Burke, John W.
  1980.   By Frank H. Winter, July 1980, telephone, notes in "Cleveland Rocket Society" file, National Air and Space Museum.
Clarke, Arthur C.
  1977.   By Frank H. Winter, 20 October 1977, Washington, D.C., notes in author's collection.
Cleater, P. E.
  1978.   By Frank H. Winter, 7 October 1978, Washington, D.C., tape in author's collection.
Delaney, Steven
  1979.   By Frank H. Winter, 9 February 1979, Alexandria, Virginia, notes in author's collection.
Ehricke, Krafft
  1978.   By Frank H. Winter, 30 June 1978, telephone, notes in "Krafft Ehricke" file, National Air and Space Museum.
Gates, Franklin M.
  1979.   By Frank H. Winter, 12 January 1979, telephone, notes in "U.S. Rocketry, 1930's, General" file, National Air and Space Museum.
von Khuon, Ernst
  1977.   By Frank H. Winter, 9 February 1977, Washington, D.C. notes in "German Rocketry, 1930s, General" file, National Air and Space Museum.
Lasser, David
  1980.   By Frank H. Winter, 29 September 1980, telephone, notes in "David Lasser" file, National Air and Space Museum.
Lemkin, William
  1977.   By Frank H. Winter, 15 October 1977, New York City, notes in "William Lemkin" file, National Air and Space Museum.
Loebell, Ernst
  1978.   By Frank H. Winter, 22 September 1978, telephone, notes in "Ernst Loebell" file, National Air and Space Museum.
Moskowitz, Sam
  1977.   By Frank H. Winter, 16 October 1977, Newark, New Jersey, notes in "U.S. Rocketry, 1930s, General" file, National Air and Space Museum.
Pendray, G. Edward
  1977a. By Frank H. Winter, 30 January 1977, Washington, D.C., notes in "G. Edward Pendray" file, National Air and Space Museum.
  1977b. By Frank H. Winter, 15 October 1977, telephone, notes, in "G. Edward Pendray" file, National Air and Space Museum.
Reeb, William F.
  1978.   By Frank H. Winter, 15 September 1978, telephone, notes in "Ernst Loebell" file, National Air and Space Museum.
Schaefer, Herbert
  1977.   By Frank H. Winter, 10 January 1977, Washington, D.C., notes in author's collection.
Settle, Admiral Thomas G. W.
  1977.   By Frank H. Winter, 27 October 1977, telephone, notes, in "Thomas G. W. Settle" file, National Air and Space Museum.
Shesta, John
  1977.   By Frank H. Winter, 18 February 1977, telephone, notes in "John Shesta" file, National Air and Space Museum.
Smith, Bernard
  1977.   By Frank H. Winter and Tom D. Crouch, 1 February 1977, Washington, D.C., notes in "Bernard Smith" file, National Air and Space Museum.
Truax, Robert C.
  1978.   By Frank H. Winter, 8 August 1978, Washington, D.C., notes in author's collection.
Weisinger, Mort
  1977.   By Frank H. Winter, 8 November 1977, telephone, notes in author's collection.
Williamson, Merritt A.
  1979.   By Frank H. Winter, 23 January 1979, telephone, notes in "U.S. Rocketry, 1930s, General" file. National Air and Space Museum.

B. Speech

Tokaty-Tokaev, Grigori Aleksandrovich
    1968.   To British Interplanetary Society, London, May 1968. In Michael Kapp Tape Collection, National Air and Space Museum, Washington, D.C.

C. Letters

Ananoff, Alexandre
    1938.   To Willy Ley, 20 October 1938, "Alexandre Ananoff" file, Willy Ley Collection, National Air and Space Museum.
Battelle Columbus Laboratories
    1979.   To Frank H. Winter, 25 January 1979, in "Lester D. Woodford" file, National Air and Space Museum.
Browne, Howard
    1978.   To Frank H. Winter, 18 March 1978, "German Rocket Society" file, National Air and Space Museum.
Brandt, C. A.
    1930.   To C. P. Mason, 10 June 1930, Pendray Papers.
Cleator, P.E.
    1931.   To C. P. Mason, 10 August 1931, Pendray Papers.
    1934a. To G. Edward Pendray, 11 April 1934, Pendray Papers.
    1934b. To G. Edward Pendray, 31 May 1934, Pendray Papers.
    1934c. To G. Edward Pendray, 28 July 1934, Pendray Papers.
    1934d. To G. Edward Pendray, 30—31 October 1934, Pendray Papers.
    1935.   To G. Edward Pendray, 7 February 1935, Pendray Papers.
    1936a. To G. Edward Pendray, 10 March 1936, Pendray Papers.
    1936b. To G. Edward Pendray, 3 April 1936, Pendray Papers.
    1936c. To G. Edward Pendray, 8 May 1936, Pendray Papers.
    1936d. To G. Edward Pendray, 14 November 1936, Pendray Papers.
    1939.   To G. Edward Pendray, 14 September 1939, Pendray Papers.
    1941.   To G. Edward Pendray, 3 April 1941, Pendray Papers.
    1963.   To G. Edward Pendray, 13 December 1963, Pendray Papers.
    1978a. To Frank H. Winter, 11 January 1978, "P. E. Cleator" file and "British Interplanetary Society" file, National Air and Space Museum.
    1978b. To Frank H. Winter, 23 October 1978, "P. E. Cleator" file and "British Interplanetary Society" file, National Air and Space Museum.
Cleveland Chamber of Commerce
    1933.   To Ernst Loebell, ca. 1933, in "Ernst Loebell" file, National Air and Space Museum.
Cleveland Public Library
    1978.   To Frank H. Winter, 20 July 1978, "Cleveland Rocket Society" file, National Air and Space Museum.
Constantinescu, Clinton
    1931.   To Nathan Schachner, 17 August 1931, Pendray Papers.
Deisch, Noel
    1931.   To David Lasser, 1 March 1931, "Noel Deisch" file, National Air and Space Museum.
von Dickhuth-Harrach, Hans Wolf and Willy Ley
    1934.   To VfR members, 4 January 1934, Herbert Schaefer collection, copy in "German Rocket Society" file, National Air and Space Museum.
Falkenberg, B.
    1931.   To Secretary of American Interplanetary Society, n.d., received 9 December 1931, Pendray Papers.
Fierst, A. L.
    1977.   To Frank H. Winter, 31 March 1977, "American Rocket Society" file, National Air and Space Museum.
Gatland, Kenneth
    1977.   To Frank H. Winter, 9 December 1977, in "British Interplanetary Society" file, National Air and Space Museum.
Hanna, Edward L.
    1964a. To John Tascher, 21 June 1964, "Cleveland Rocket Society" file, National Air and Space Museum.
    1964b. To John Tascher, 23 September 1964, "Cleveland Rocket Society" file, National Air and Space Museum.
Hasegawa, Tatsue
    1935.   To Editor, *Astronautics,* 17 July, "Japan, 1930—1945" file, National Air and Space Museum
Kaiser, Hans K.
    1939.   To Ernst Loebell, 2 August 1939, "Hans K. Kaiser" file, National Air and Space Museum.
Klemin, Dr. Alexander
    1934.   To G. Edward Pendray, 16 March 1934, Pendray Papers.
Koizumu, I.

1931a. To G. Edward Pendray, 25 August 1931, Pendray Papers.

1931b. To G. Edward Pendray, 10 September 1931, Pendray Papers.

de Koningh, Gerrit

1948. To Eric Burgess, 6 October 1948, "Eric Burgess" file, National Air and Space Museum.

Lasser, David

1977. To Frank H. Winter, 14 April 1977, "David Lasser" file, and "American Rocket Society" file, National Air and Space Museum.

Lencement, Robert

1937. To Ernst Loebell, 27 July 1937, "Ernst Loebell" file, National Air and Space Museum.

1938. To Ernst Loebell, 14 November 1938, "Ernst Loebell" file, National Air and Space Museum.

Ley, Willy

1931a. To G. Edward Pendray, 23 July 1931, Pendray Papers.

1931b. To G. Edward Pendray, 6 October 1931, Pendray Papers.

1931c. To G. Edward Pendray, 2 November 1931, Pendray Papers.

1932a. To G. Edward Pendray, 5 January 1932, Pendray Papers.

1932b. To G. Edward Pendray, 30 June 1932 (post card), Pendray Papers.

1933. To American Interplanetary Society, 26 December 1933, Pendray Papers.

1934. To G. Edward Pendray, 2 February 1934, Pendray Papers.

1935. To G. Edward Pendray, 7 February 1935, Pendray Papers.

1952. To Edward Peck, 14 January 1952, "Ernst Loebell" file, National Air and Space Museum.

Loebell, Ernst

1978a. To Frank H. Winter, 19 October 1978, "Ernst Loebell" file and "Cleveland Rocket Society" file, National Air and Space Museum.

1978b. To Frank H. Winter, 10 December 1978, "Ernst Loebell" file, National Air and Space Museum.

1979. To Frank H. Winter, 31 January 1979, "Ernst Loebell" file, National Air and Space Museum.

Malina, Frank

1936-
1946
To F. Malina family, "Rocket Research and Development—Excerpts from Letters Written Home by Frank J. Malina Between 1936 and 1946," unpublished monograph, National Aeronautics and Space Administration Historical Archives, Washington, D.C.; also "Frank Malina" file, National Air and Space Museum.

Mansell, J.

1931. To G. Edward Pendray, 16 August 1931, Pendray Papers.

Matarazzo, Ezio

1934. To Herbert Schaefer, 24 March 1934, Herbert Schaefer collection, copy in "Rocketry, Astronautics, General, 1930s" file, National Air and Space Museum.

1935. To Herbert Schaefer, 4 February 1935, Herbert Schaefer collection, copy in "Rocketry, Astronautics, General, 1930s' file, National Air and Space Museum.

Moskowitz, Sam

1974. To Frank H. Winter, 22 November 1974, "American Rocket Society" file, National Air and Space Museum.

1977. To Frank H. Winter, 6 October 1977, "American Rocket Society" file, National Air and Space Museum.

Palmer, Marjorie

1977. To Frank H. Winter, 6 December 1977, "German Rocket Society" file, National Air and Space Museum.

1978. To Frank H. Winter, 18 March 1978, "German Rocket Society" file, National Air and Space Museum.

Pendray, G. Edward

1932. To Dr. Samuel Lichenstein, 31 October 1932, Pendray Papers.

1934a. To Laurence Manning, 13 February 1934, Pendray Papers.

1934b. To Harry W. Bull, 3 October 1934, Pendray Papers

1938. To James H. Wyld, 28 June 1938, Pendray Papers.

1939. To Morton Savell, 4 January 1939, Pendray Papers.

1976a. To Frank H. Winter, 30 January 1976, "G. Edward Pendray" file and copy in "American Rocket Society" file, National Air and Space Museum.

1976b. To Frank H. Winter, 6 May 1976, "G. Edward Pendray" file and copy in "American Rocket Society" file, National Air and Space Museum.

1978. To Frank H. Winter, 28 March 1978, "G. Edward Pendray" file and copy in "American Rocket Society" file, National Air and Space Museum.

Pierce, H. F.

1931. To G. Edward Pendray, 27 October 1931, Pendray Papers.

Poggensee, Karl

1977. To Frank H. Winter, 1 March 1977, "Karl Poggensee" file and copy in "Ernst Loebell" file, National Air and Space Museum.

Prindle, Jr., Charles A.

1934. To G. Edward Pendray, 6 June 1934, Pendray Papers.

Riedel, Babs

1977. To Howard Wolko, 5 September 1977, "Klaus Riedel" file, National Air and Space Museum.

Sänger-Bredt, Dr. Irene
  1977. To Frank H. Winter, 29 August 1977, "Irene Sänger-Bredt" file, National Air and Space Museum.
Schaefer, Herbert
  1936. To Ezio Matarazzo, 7 August 1936, Herbert Schaefer collection and copy in "Rocketry, Astro-
          nautics, General, 1930s" file, National Air and Space Museum.
Schiff, Judith A.
  1978. To Frank H. Winter, 10 November 1978, "Yale Rocket Club" file, National Air and Space Museum.
Schlesinger, W. L.
  1955. To Andrew G. Haley, 21 April 1955, Andrew G. Haley Collection, National Air and Space Museum.
Schofield, Robert E.
  1965. To Ernst Loebell, 29 November 1965, "Cleveland Rocket Society" file, National Air and Space
          Museum.
Sykora, Fritz
  1980. To Frank H. Winter, 28 August 1980, "Austrian Rocket Society" file, National Air and Space
          Museum.
Tabanera, Teófilo
  1979a. To Frank H. Winter, 24 February 1979, "Teófilo Tabanera" file, National Air and Space Museum.
  1979b. To Frank H. Winter, 2 April 1979, "Teófilo Tabanera" file, National Air and Space Museum.
Thoolen, A. G.
  1935. To G. Edward Pendray, n.d. 1935, Pendray Papers.
Van Devander, Charles
  1976. To Frank H. Winter, 20 April 1976, "American Rocket Society" file, National Air and Space
          Museum.
VfR
  1931. To C. S, Henshaw, 6 October 1931, Pendray Papers.
Zoike, Helmut
  1978. To Frank H. Winter, 26 December 1978, "Helmut Zoike" file, National Air and Space Museum.

D. Unpublished Material

Acme Newspictures photo J280464, Tsunedo Obara's Rocket
  1934. In "Rocketry, 1930s Gen." file, National Air and Space Museum. Also published (see Published
          Sources, "From Tokyo," "A Japanese," and, Photograph, 1935).
American Interplanetary Society
  1930ca."Membership List-Active," ca. 1930, Pendray Papers.
  1931. "Supplementary Mailing List for Bulletin," ca. 1931, Pendray Papers.
American Museum of Natural History
  1931. Bill of hall rental to American Interplanetary Society, 7 February 1931, Pendray Papers.
American Rocket Society
  1934. "Draft of a Proposal for the Establishment of a Fund for Rocket Research with the Object of
          Developing High-Altitude Rockets for Scientific and Meteorological Investigation," ca. 1934,
          Pendray Papers.
Corn, Joseph
  1977. "A New Sign in the Heavens: The Gospel of Aviation and American Society, 1880-1950." Ph.D.
          Thesis, University of California, Berkeley.
Exposition International des Arts et des Techniques, 1937
  1937. "Diplôme Commémoratif." Issued to Ernst Loebell, photostat in "Ernst Loebell" file, National Air
          and Space Museum.
Haley, Andrew G.
  n.d.   "Some Capsule Notes on Rocket-Astronautical Societies." In folder, "Chapter VIII—The Founding
          Societies," file box 19, Andrew G. Haley Papers.
Heinisch, Kurt
  1946. Interrogation of, in U.S. Army Air Corps, Intelligence Div., Appendix to Developments of
          Peenemünde East, Wright Field Document D 72.31/48 Part 1 of 3 parts, declassified 8 July 1946.
Helfers, M. C.
  1954. Employment of the V-Weapons, Office of the Chief of Ordnance, Army Historical Office,
          Washington, D.C., 31 May 1954, unpublished paper.
Houtman, Maarten
  1930s. Dutch astronautical literature, 1930s private collection on loan to the National Air and Space
          Museum.
"Lasser, David"
  n.d.   Biographical file, National Air and Space Museum.
"Lear, William P."
  n.d.   Biographical file, National Air and Space Museum.
"Lichtenstein, Samuel"
  n.d.   Biographical file, National Air and Space Museum.
Manning, Laurence
  1931. "American Interplanetary Society, Inc. Financial Statement," Pendray Papers.

Mason, C. P.
   1931. "Report of the Secretary, 1930—1931." Pendray Papers.
Mendillo, Michael and David DeVorkin
   1977. "The History of the Martian Canals," preprint, paper presented at the 1977 Annual Meeting of the American Association for the Advancement of Science (AAAS), 20—25 February 1977, Denver, Colorado.
Nebel, Rudolf
   n.d.   "Raketen-Artillerie." Unpublished mimeograph report in "Verein für Raumschiffahrt" file, Willy Ley collection, National Air and Space Museum.
"Pierce, Hugh Franklin"
   n.d.   Biographical file, National Air and Space Museum.
Pratt, Fletcher
   1930— "American Interplanetary Society Librarian's Report, 1930—31," Pendray Papers.
   31.
"Pratt, Fletcher"
   n.d.   Biographical file, National Air and Space Museum.
"Püllenberg, Albert"
   n.d.   Biographical file, National Air and Space Museum.
"REP-Hirsch Prize"
   n.d.   File in the National Air and Space Museum.
Ritchie, Donald J.
   1966.   Rocket and Missile Systems Development in the Soviet Union. Southfield, Michigan. Unpublished. Copy in National Air and Space Museum.
"Rynin, Nikolai Alexyevich"
   n.d.   Biographical file, National Air and Space Museum.
Schaefer, Herbert
   1933-   Private Notebook. Private collection of Herbert Schaefer, San Diego, California.
   1936.
Shesta, John
   1978.   "Reaction Motors Incorporated—First Large Scale American Rocket Company: A Memoir." Edited by Frank H. Winter, paper presented at the 29th Congress, International Astronautical Federation, Dubrovnik, Yugoslavia, 1-8 October 1978. Preprint in "John Shesta" file, National Air and Space Museum.
Schulz, Werner
   1979.   "Walter Hohmann's Contributions Towards Space Flight: an Appreciation on the Occasion of the Centenary of his Birthday," preprint, paper presented at the 30th Congress, International Astronautical Federation, Munich, 17-22 September 1979.
"Verein für Raumschiffahrt"
   n.d.   File in Willy Ley collection, National Air and Space Museum.
"Welsh, Ernest"
   n.d.   Biographical file, National Air and Space Museum.
Winter, Frank H.
   1976.   "Harry Bull, American Rocket Pioneer." paper presented at the 27th Congress, International Astronautical Federation Anaheim, California, 10-17 October 1976. Preprint in "Harry Bull" file, National Air and Space Museum.
Wyld, James H.
   1938.   "A Program of Research in Experimental Rocketry," 28 June 1938, Pendray Papers.
Sources for Letters and Other Unpublished Material

Haley Papers, Andrew G., National Air and Space Museum, Washington, D.C.
Ley Papers, Willy. National Air and Space Museum, Washington, D.C.
Pendray Papers, G. Edward. Seeley G. Mudd Manuscript Library, Princeton University, Princeton, New Jersey.
Schaefer Papers, Herbert. Private collection of Herbert Schaefer, San Diego, California.

E. Published Sources

"Advances in Travel"
   1928.   Science and Invention, 16 (May):23.
Advertisement
   1932.   Brooklyn Daily Eagle, 1 March. G. Edward Pendray radio-television talk.
   1934.   The Airpost Journal, 5 (May):21. Cleveland Rocket Society, rocket flight covers.
   1937.   Journal of the British Interplanetary Society. 4 (February):5.
Africano, Alfred
   1936.   "Report on Rocket Motor Tests of August 25th." Astronautics, 33 (March):3—5, 20.
Aldiss, Brian W.
   1973.   Billion Year Spree—The True History of Science Fiction. New York: Schocken Books.

"Alexandre Ananoff"
  1950. *Weltraumfahrt,* 2 (April):41-42.
"All Aboard for Venus"
  1928. *Air Travel News,* 1 (April):19.
Allward, M. F., L. J. Carter, H. E. Ross, and K. W. Gatland
  1967a. "Astronautics in Britain, 1." *Spaceflight.* 9 (May):150-152.
  1967b. "Astronautics in Britain, 2." *Spaceflight.* 9 (June):201-206.
  1967c. "Astronautics in Britain, 3." *Spaceflight,* 9 (July):234—236.
"Altitude Rocket Explodes"
  1931. *Bulletin of the American Interplanetary Society,* 7 (February):8.
The American Interplanetary Society
  1930- *Constitution American Interplanetary Society,* mimeographed, with amendments to 1935, made
  35.   available to AIS—ARS members.
"The American Interplanetary Society"
  1931. *Wonder Stories,* 3 (June):134.
"Amtlich Bekanntmachung"
  1928. *Die Rakete,* 3 (June):134.
"Analecta"
  1935a. *Journal of the British Interplanetary Society,* 2 (May):5.
  1935b. *Journal of the British Interplanetary Society,* 2 (October):13.
Ananoff, A.
  1931. *"Éclipse Totale de Lune du 26 Septembre 1931." Bulletin de la Société Astronomique de France,* 45 (October):463—466.
  1933. *"Le Planeté Mars en 1933 Travaux Effectués a L'Observatoire de la Société." Bulletin de la Société Astronomique de France,* 47 (September):409—410.
  1935. *"La Navigation Interplanétaire—Fusées Sondes et Navires-Fusées." Bulletin de la Société Astronomique de France,* 49 (September):419—437.
  1937a. *"Les Experiences sur les Fusées a Travers le Monde." Bulletin de la Société Astronomique de France,* 51 (May):225—228.
  1937b. *"Les Expériences sur les Fusées a Travers de Monde." Bulletin de la Société Astronomique de France,* 51 (June):271—278.
  1940. "Rocketry in France." *Astronautics,* 47 (November):3—6.
  1969. *Astronautics,* NASA TT-F-12,220, Washington, D.C.: National Aeronautics and Space Administration.
  1978. *Les Mémoires d'un Astronaute ou L'Astronautique Française,* Paris: Albert Blanchard.
Andres, Ugo (pseudonym of G. Edward Pendray)
  1934. "Men of Space." *New Outlook,* 164 (October):26—33.
"Another Vice-President for the Society."
  1936. *Journal of the British Interplanetary Society.* 3 (June):22, 34.
"Anti-Aircraft Rocket Carries Machine Guns."
  1935. *Popular Science Monthly,* 126 (January):42.
Astashenkov, P. T.
  1971. *Academician S. P. Korolev, Biography* (Dayton: Air Force Systems Command, Foreign Technology Division), translation FTD-HC-23-542-70, 3 March 1971.
"L'Astronomie a L'Exposition"
  1937. *Bulletin de la Société Astronomique de France.* 51 (April):185—188.
"Attend Settle-Fordney Banquet"
  1933. *Space,* mimeograph bulletin of the Cleveland Rocket Society, 1 (December):2.
"Auf Wiedersehen"
  1933. *Space,* mimeograph bulletin of the Cleveland Rocket Society, 1 (December):b.
"Aus der Gesellschaft"
  1968. *Astronautik,* 5 (August—September—October):131—132.
"Austria"
  1934. *Journal of the British Interplanetary Society,* 1 (April):17.
Bainbridge, William Simms
  1976. *The Spaceflight Revolution—A Sociological Study.* New York: John Wiley and Sons.
v. B. "Rakettenpostzwendel"
  1935. *Nederlansch Maandblad voor Philatelie,* 14 (June):124—127.
"Bekanntmachung"
  1927. *Die Rakete,* 1 (November:142.
Bergaust, Erik
  1976. *Wernher von Braun,* Washington, D.C.: National Space Institute.
"Bibliographie—Les autres mondes sont-ils habités?"
  1913. *Cosmos,* 62 (March):277.
"Biography of Trevor Cusack Founder and Secretary of M.A.A."
  1941. *Spacewards Journal of the Manchester Astronautical Association,* 2 (January):3—6.
Blagonravov, Anatoli
  1968. "Spaceship Designer Sergei Korolev." *Space World,* E-7-55 (July):9—11.

Bloom, Ursula
    1958. *He Lit The Lamp—A Biography of Professor A. M. Low.* London: Burke Publishing Co., Ltd.
Blosset, Lise
    1974. "Robert Esnault-Pelterie: Space Pioneer." In Frederick C. Durant, III and George S. James, editors, *First Steps Towards Space*—Smithsonian Annals of Flight, 10:5—31. Washington, D.C.: Smithsonian Institution Press.
"Boy of 16 Tells of Plans to Reach Moon"
    1937. *Daily Mail* (London), 4 February 1937.
"Branch Club News"
    1935. *The International Observer,* mimeographed newsletter of the International Cosmos Science Club, New York, Sam Moskowitz collection, 1 (May):1.
von Braun, Wernher
    1930— Lunetta, *Leben und Arbeit* (Berlin), 2 & 3:89—92.
    1931.
    1932. "*Das Geheimnis der Flüssigkeitsrakete,*" *Umschau* 36 June:449—452.
    1950. "Correspondence Prof. Hermann Oberth—A reply to Otto Hester," *Journal of the British Interplanetary Society,* 9 (March):87—89.
    1967. "German Rocketry." In Arthur C. Clarke, editor, *The Coming of the Space Age:* 33—35. New York: Meredith Press.
von Braun, Wernher, and Frederick I. Ordway, III
    1969. *History of Rocketry and Space Travel.* New York: Thomas Y. Crowell Company.
"British Rocket Car Experiment Interplanetary Society Designing Mobile Test Bed."
    1934. *The Autocar,* 72 (March):347.
"British Rocket Mail Trials"
    1934. *Airpost Journal,* 5 (August):8.
"British Scientists Making for Mars"
    1934. *Sunday Graphic* (London) 4 February 1934.
"Briton Is Making 'Torpedo Rocket' to Entangle and Bomb Foe's Planes"
    1938. *New York Times.* (June):24.
Brügel, Werner
    1933. *Männer der Rakete.* Leipzig: Verlag Hachmeister and Thal.
    1936. "I.R.K.A. An International Bureau of Information on Rocketry." *Journal of the British Interplanetary Society,* 3 (February):6.
Bullock, Alan
    1962. *Hitler—A Study in Tyranny.* New York: Harper Torchbooks.
Burgess, Eric
    1944. "Rocket Experiments in Manchester." *Astronautics,* 59 (September):6—10.
"By Rocket to the Moon!
    1936. *The Ashton Reporter.* (Ashton Under Lyne, Lancashire, England), 16 October 1936.
"By Rocket to the Moon Shown At College Auditorium"
    1933. *Space,* mimeograph bulletin of the Cleveland Rocket Society, 1 (December):1.
"California Rocket Society"
    1943. *Astronautics,* 56 (December):13.
"Centerbladet"
    1937. *Journal of the British Interplanetary Society,* 4 (February):13.
Cleator, P. E.
    1933. "British 'Rocketeers—Reaching the Moon—And Elsewhere." *Liverpool Echo,* 8 September 1933.
    1934a. "Retrospect and Prospect." *Journal of the British Interplanetary Society,* 1 (January):2—4.
    1934b. "Editorial—The Attitude of the Government," *Journal of the British Interplanetary Society.* 1 (April):13—15.
    1934c. "Editorial—Experimenters," *Journal of the British Interplanetary Society,* 1 (April):13.
    1934d. "Editorial—Experimental Work in England." *Journal of the British Interplanetary Society,* 1 (April):14.
    1934e. "The Press: An Appreciation—And A Plea." *Journal of the British Interplanetary Society,* 1 (April):20.
    1934f. "Herr Zucker's Postal Rocket." *Journal of the British Interplanetary Society,* 1 (July):27.
    1935. "Editorial." *Journal of the British Interplanetary Society,* 2 (May):2.
    1936a. "Editorial." *Journal of the British Interplanetary Society,* 3 (February):2.
    1936b. "Editorial," *Journal of the British Interplanetary Society,* 3 (June):15—16.
    1936c. *Rockets Through Space,* New York: Simon and Schuster.
    1938a. "The Rocket Ban," *The Astronaut,* Manchester Interplanetary Society, 2 (February):3—6.
    1938b. "Hymn to Progress." *The Astronaut,* Manchester Interplanetary Society, 2 (August):15—18.
    1948. "Autopsia." *Journal of the British Interplanetary Society,* 7 (May):97—99.
    1950. "Matters of No Moment." *Journal of the British Interplanetary Society,* 7 (May):97—99.
    1961. *An Introduction to Space Travel.* New York: Pittman Publishing Corporation.
Cleveland City Directory 1930
    1930. Cleveland: The Cleveland City Directory Co.
Cleveland City Directory 1935

1935. Cleveland: The Cleveland City Directory Co.

"Cleveland Experiments"
1936. *Journal of the British Interplanetary Society,* 2 (June):25.

"The Cleveland Rocket Society Extends Appreciation To . . ."
1933. *Space,* 1 (December):c. Mimeographed bulletin of the Cleveland Rocket Society.

"Cleveland Gets Races For Next Five Years"
1935. *New York Herald Tribune,* 29 September 1935.

"Cleveland Rocket Society Sends Model Rocket to Paris International Exhibition"
1937. *The Cleveland Plain Dealer,* 11 April 1937.

Clifton New Jersey City Directory 1956
1956. Newark, J. J.: The Price & Lee Co.

"Communication Écrites"
1942. *Bulletin de la Société Astronomique de France,* 56 (January):12—15.
1945. *Bulletin de la Société Astronomique de France.* 59 (January—February—March):17—21.

"Complimentary Membership Conferred"
1934. *Space,* mimeographed bulletin of the Cleveland Rocket Society, 5 (September):12.

"Constitution of the Cleveland Rocket Society"
1933. *Space,* mimeographed bulletin of the Cleveland Rocket Society, 1 (December):4—7.

" 'The Conquest of Space' Coming"
1931. *Bulletin of the American Interplanetary Society,* 11 (August):10.

"Correspondence—A London Section?"
1936. *Journal of the British Interplanetary Society,* 3 (June):32.

Cranz, Carl, et al. .
1926. *Lehrbuch der Ballistik.* Berlin: Verlag von Julius Springer.

"CRS Rocket Pins"
1933. *Space,* mimeographed bulletin of the Cleveland Rocket Society, 1 (December):2.

Davis, Douglas
1939. "Rocket Metals," *Journal of the Peoria Rocket Society,* 1 (January):2.

Dawson, P. J., Compiler
1948. *German Organisations and Personalities Engaged in Research and Development of Armament During the Second World War,* (London): Technical Information Bureau for Chief Scientist, Ministry of Supply, October 1948.

"Death-Dealing Rocket"
1924. *Science And Invention,* 12 (November):656.

"Death Ray Matthews"
1973. *South Wales Echo,* (Cardiff) 22 May 1973.

"Death Rocket Rains Fiery Metal"
1924. *Popular Science Monthly,* 105 (November):40.

von Dickuth-Harrach, Maj., [Hans-Wolf]
1928. "Probleme der Mondfahrt." *Berliner Tageblatt,* 8 February 1928.
1937. "Der Untergang des Luftschiffes 'Hindenburg.' " *Das Neue Fahrzeug,* 4 (May):2—4.

"Die Tatigkeit der Gasellschaft im Jahre 1937."
1938. *Mitteilungen der Gesellschaft für Weltraumforschung Beilage Nr. 1.* In *Astronomische Rundschau,* 1 (October):I—III.

Dooling, David
1974. "The Evolution of the Apollo Spacecraft." *Spaceflight,* 16 (March:82—88)

Dornberger, Walter R.
1954. "European Rocketry After World War I." *Journal of the British Interplanetary Society,* 13 (September):245—262.
1958. *V-2.* New York: The Viking Press.
1963. "The German V-2." *Technology and Culture,* 4 (Fall):393—409.

"Dr. Lyon Selects Desert for Rocket Experiment"
1931. *Bulletin of the American Interplanetary Society,* 10 (June—July):3.

Drummond-Hay, Lady [Marguerite]
1930. "Gossip of the World." *The Mentor-World Traveller,* 22 (October):24—26.

Durant, III, F. C.
1961. *The International Astronautical Federation 1960—61.* (Wilmington, Mass.: Research and Development Division, Avco Corporation).

"Editorial"
1936. *Journal of the British Interplanetary Society,* 3 (June):15—16.
1937a. *Journal of the Brixtish Interplanetary Society,* 4 (February):3—4.
1937b. *Journal of the British Interplanetary Society,* 4 (December):1.

"Ein Brief aus Afrika"
1929. *Die Rakete,* 3 (March):43—44.

"Ein Jahr Verein für Raumaschiffahrt E. V."
1928. *Die Rakete,* 2 (July):99—100.

Ellington, Jesse T., and Perry F. Zwisler
1967. *Ellington-Zwisler Rocket Mail Catalog.* New York: John W. Nicklin.

"England and Science Fiction"
   1931. *Wonder Stories*, 2 (March):1195.
"*Errichtung einer Geschaftsstelle des Vereins für Raumschiffahrt E. V. in Berlin*"
   1929. *Die Rakete*, 3 (August):9.
"*Esposizione Aeronautica Vienese*"
   1928. *Aeronautica*, (Milan) 2 April 1928:308.
Essers, I.
   1976. *Max Valier—A Pioneer of Space Travel*, NASA TT F-664, Washington, D.C.: National Aeronautics
          and Space Administration.
"Experiments in America"
   1931. *Bulletin of the American Interplanetary Society*, 14 (December):8.
"Explosive Mixture They Will Not Use"
   1937. *The Manchester Evening News*, 14 June 1937.
Fedorov, A. S.
   1970. "K. E. Tsiolkovsky As A Film Consultant." Translation of "*K. E. Tsiolkovsky Konsul'tieuyet Film.*"
          Piroda. Moscow: USSR Academy of Sciences, 5 (May):1. NASA Technical Translation TT F-14,890.
[Field, Francis, J.]
   1937. " 'NG' and 'NT.' " *The Aero Field.* 1 (January—February):19.
"5000 RM. Werbprämien"
   1927. *Die Rakete*, 15 November 1927:115.
"The Flight of Experimental Rocket No. 2"
   1933. *Astronautics*, 26 (May):1—11.
"The Flight of Rocket No. 4"
   1934. *Astronautics*, 30 (October—November):1—2, 12.
"Flight to Venus Not Started"
   1928. *Riverside Daily Press.* (Riverside, California) 5 March 1928.
"*Flug Rundschau*"
   1936. *Flugsport*, 28 (March):79—82.
"The Forthcoming Annual Meeting"
   1934. *Astronautics*, 30 (October—November):7.
"*Fortschrittliche Verltehrstechnik E. V.*"
   1934. *Das Neue Fahrzeug*, 2 (April):16.
"The Four Winds Items of Interest From All Quarters"
   1934. *Flight*, 26 (December):1376.
Freedom, Russel
   1965. *Jules Verne Portrait of A Prophet.* New York: Holiday House.
"Friedrich Wilhelm Sander"
   1968. *Astronautik*, 4:131.
Fritz, A.
   1966. "Guido von Pirquet," *Mitteilungen der DGRR Astronautische Berichte*, 19 (May):23.
"From Tokyo, Japan" (picture)
   1935. *Astronautics*, 31 (June):2.
"From the World's Scrap-Book: New Items of Topical Interest"
   1928.. *The Illustrated London News*, 172 (March):356.
"The Future of the Rocket"
   1931. *Bulletin of the American Interplanetary Society*, 8 (March—April):6.
Gail, Otto Willi
   1928. *Raket Genom Världsrymed.* Stockholm: Tryckeri Aktiebolaget.
Gartmann, Heinz
   1956. *The Men Behind the Space Rockets.* New York: David McKay Co.
Gates, Franklin M. and Merritt A. Williamson
   1937. "Astronautics, A New Science." *The Yale Scientific Magazine*, 11 (Winter):3, 20—21.
"*Gedenken an. Dipl.-Ing. Rudolf Nebel*"
   1978. *Astronautik*, 4:123—124.
Generales, Jr., Constantine D. J.
   1974. "Recollections of Early Biomedical Moon-Mice Investigations." In Frederick C. Durant, III and
          George S. James, *First Steps Toward Space*—Smithsonian Annals of Flight 10:75—80. Washing-
          ton, D.C.; Smithsonian Institution Press.
"*Gerhardt Zucker, Raketenerfinder*"
   1936. *Flugsport*, 28 (February):80.
"German Society Busy"
   1930. *Bulletin of the American Interplanetary Society*, 5 (November-December):4.
Gifford, Dennis
   1971. *Science Fiction Film.* London: Studio Vista.
Giles, Cedric
   1944. "The Rocket Societies." *Astronautics*, 60 (December):12—13.
"Glushko, Valentin Petrovich."
   1968. In Edward L. Crowley, et al., editors. *Prominent Personalities in the USSR:* 181. Metuchen, N.J.:
          Scarecrow Press.

Glushko, V. P.

1973. *Development of Rocketry and Space Technology in the USSR*. Moscow: USSR Academy of Science Novosti Press Publishing House.

1974. "The Leningrad Gas Dynamics Laboratory (GDL) Contribution to the Development of Rocketry." In *Proceedings of the Thirteenth Congress of the History of Science—Section XII—The History of Aircraft, Rocket and Technology:* 30—40. Moscow: *Izdatelstvo "Nauka."*

1977. *Puta v Raketnoi Tekhniki Izbranniye Trudi 1924—1946.* Moscow: Mashinostroenni.

Goddard, Esther C. and G. Edward Pendray, editors.

1970. *The Papers of Robert H. Goddard.* New York: McGraw-Hill Book Co.

Goddard, Robert H.

1919. *A Method of Reaching Extreme Altitudes.* Washington, D.C.: Smithsonian Institution Press.

Golovanov, Yaroslav

1975. *Sergei Korolev—The Apprenticeship of A Space Pioneer.* Moscow: MIR Publishers.

"Good Day! Mr. Mars, said Prof. Todd"

1919. *Electrical Experimenter,* 10 (October):525.

Goudket, Herbert E.

1935. "The Astronomer." *The International Observer,* mimeographed newsletter of the International Cosmos Science Club, New York, Sam Moskowitz Collection, 1 (May):14—15.

"Gráve, Alexander Dmitri"

1972. In *Bol'shaia Sovetskaia Entsiklopediia,* 7:197, Moscow: *Izdatelstvo "Sovetskaia Entsiklopediia."*

Great Britain

1875. *The Public General Acts Passed in the Thirty-Eight and Thirty-Ninth Years in the Reign of Her Majesty Queen Victoria.* London: George and Ward Eyre and William Spottiswoode.

"Guido von Pirquet"

1966. *Mitteilungen der DGRR Astronautische Berichte.* 19 (May):23.

Hagedorn and Co.

1978. *51. Spezialauktion 27./28.10. 1978.* Limburg, West Germany: Hagedorn and Co.

Haley, Andrew G.

1958. *Rocketry and Space Exploration.* Princeton, New Jersey: D. Van Nostrand Co., Inc.

Hamon, A.

1931. "L'Éclipse Totale de Lune du 2 Avril 1931." *Bulletin de la Société Astronomique de France,* 45 (May:205—206.

Hanna, Edward L.

1933a. "An Idea Becomes Reality." *Space,* mimeographed bulletin of the Cleveland Rocket Society, 1 (December):2, 8.

1933b. "Space," *Space,* mimeographed bulletin of the Cleveland Rocket Society, 1 (December):a.

Hanson, Maurice K.

1939. "The Payload of the Lunar Trip." *Journal of the British Interplanetary Society,* 5 (January):10—16.

Harper, Harry

1946. *Dawn of the Space Age,* London: Sampson Low, Marston and Co., Ltd.

Hartl, Hans

1958. *Hermann Oberth Vorkampfer der Weltraumfahrt.* Hannover: Theodor Oppermann Verlag.

"Harvey Firestone Sr. Hinted As Sponsor for Stratosphere Flight"

1934. *The Cleveland Plain Dealer,* 28 January 1934.

"Has Two Step Rocket Ready"

1931. *Bulletin of the American Interplanetary Society,* 6 (January):1.

[Healy, Roy]

1941. "Wyld Motor Retested Three Runs of August 1st" *Astronautics,* 50 (October):8.

"To Hear $2,500,000 Suit On Airplane Patent"

1931. *New York Times,* 19 September 1931.

Heflin, Woodford A.

1969. "Who Said It First? 'Astronautics' " *Aerospace Historian,* 16 (Summer):44—47.

"Herr Willy Ley, New York" (picture).

1939. *Weltraum,* 1 (April):17.

"Herr Zucker's Postal Rocket"

1934. *Journal of the British Interplanetary Society.* 1 (July):27.

Hingley, Ronald

1974. *Joseph Stalin: Man and Legend,* London: Hutchison & Co.

Hirsch, André-Louis

1930. "Les Recherchés Astronautiques a l'Etranger" *Bulletin de la Société Astronomique de France,* 44 (July):323—324.

"History In Brief"

1939. *Journal of the Peoria Rocket Association,* 1 (January):1. Mimeographed bulletin in collection of Frederick I. Ordway, III.

"Ing. Guido von Piquet"

1950. *Journal of the British Interplanetary Society,* 9 (July):204—206.

"In League with the Future"

1961. *The Open Shelf.* Cleveland Public Library, 3—6 (March-June):14.

"International Interplanetary News"

    1934.   *Journal of the British Interplanetary Society,* 1 (April):16—19.

"International Rocket Society and Bureau of Information"

    1934.   *Space,* mimeographed Journal of the Cleveland Rocket Society, 4 (August):13.

"Interplanetary Progress"

    1931.   *Wonder Stories,* 2 (April):1332.

"Interplanetary Society Progresses"

    1930.   *Wonder Stories,* 2 (December):754.

"Interesting Talks Heard on"

    1934.   *Akron-Times Press,* (Akron, Ohio), 16 January 1934.

Janser, Arthur

    1939.   "Fuels and Motors." *Journal of the British Interplanetary Society,* 5 (January):23—25.

"A Japanese Rocket"

    1935.   *Popular Aviation,* 16 (January):18.

Journal of the British Interplanetary Society, 1937

    1937.   "Editorial." 4 (February):3.

Kaiser, Hans K.

    1938.   *"Mitteilungen der 'Breslauer astronomischen Vereinigung e. V.'* Die Arbeit der B.a.V. im 1937." *Astronomische Rundschau,* 1 (January):14—15.

    1939a.   *"Die 'Section Astronautique' Frankreichs."* Weltraum, 1 (January):8—9

    1939b.   *"Raketenversuche in Scotland."* Weltraum, 1 (April):25—27.

    1949.   "The Spirit of Astronautics in Germany in the Last 15 years." *Journal of the British Interplanetary Society,* 8 (March):49—51.

    1954.   *Les Fusées Véhicules de l'Avenir.* Paris: Amiot Dumont.

Von Kármán, Theodore with Lee Edson

    1967.   *The Wind and Beyond.* Boston: Brown and Company.

"Kleimenov, Ivan Terentyevich"

    1969.   In Prof. G. V. Petrovich, editor. *The Soviet Encyclopedia of Space Flight:* 209. Moscow: Mir Publishers.

Klemin, Dr. Alexander

    1937.   "On the Aerodynamic Principles of the Greenwood Lake Rocket Aeroplane." *Astronautics,* 36 (March):7—9.

Koelle, H. H.

    1955.   "Astronautical Activities in Germany." *Journal of the British Interplanetary Society,* 14 (May-June):121—132.

Korneev, L. K., editor.

    1964.   *F. A. Tsander Problems of Flight by Jet Propulsion,* NASA TTF-147. Jerusalem: Israel Program for Scientific Translations.

"Korolev, Sergey Pavlovich"

    1973.   In Dictionary of Scientific Biography, 7 (1973):465. New York: Charles Scribner's Sons.

Kramarov, G. M.

    1962.   *Pervorev Mir Obshchestvo Kosmonavtiki.* Moscow: *Izdatelstvo "Znanye."*

Kulagin, I. I.

    1974.   "Developments in Rocket Engineering Achieved by the Gas Dynamics Laboratory in Leningrad." In Frederick C. Durant, III and George S. James, editors, *First Steps Toward Space.* Smithsonian Annals of Flight 10:91—102. Washington, D.C.: Smithsonian Institution Press.

"Kurze Berichte"

    1939a. *Weltraum,* 1 (January):14—15.

    1939b. *Weltraum,* 1 (December):49.

Kyle, David

    1976.   *A Pictorial History of Science Fiction.* London: Hamlyn Publishing Group, Ltd.

Langemak, G. E., and V. P. Glushko.

    1935.   *Rockets, Their Construction and Use.* (Translation of *Raketen, Ihre Konstruction und Verwendung,* Supreme Command of the Navy, Research Developments and Patents Branch, Berlin, 1940?), originally published in Russian by the Dept. of Scientific and Technical Information, People's Commissariat of Heavy Industry USSR, and the Main Editorial Office for Aeronautical Literature, Moscow-Leningrad, 1935). National Aeronautics and Space Administration, Washington, D.C., 1978 (NASA Technical Memorandum, NASA TM-75021).

Lasser, David

    1932.   "Report of the President of the American Interplanetary Society." *Astronautics,* 19 (May):7—8.

"Laurence E. Manning Becomes President."

    1933.   *Astronautics,* 26 (May):12—14.

"Laurence Manning: Rocket Pioneer, etc."

    1961.   *Asbury Park Sunday Press.* (New Jersey) 22 October 1961.

Lee, Asher

    1962.   *The Soviet Air Force.* New York: The John Day Co.

Lehman, Milton

    1963.   *This High Man.* New York: Farrar, Strauss, and Giroux.

Lemkin, William

1931. "Rocket Fuels." *Bulletin of the American Interplanetary Society,* 6 (January): 2—5.

Lencement, Robert

1937. *"Astronautiques at the 'Palais de la Découverte."* *Journal of the British Interplanetary Society,* 4 (December): 20—22.

"Leningrad Group for Study of Reactive Motion"

1969. In Prof. G. V. Petrovich, editor, *The Soviet Encyclopedia of Space Flight:* 217—218. Moscow: Mir Publishers.

"A Letter from Dr. Goddard"

1931. *Bulletin of the American Interplanetary Society.* 10 (June-July): 9—10.

"A Letter from the German Interplanetary Society"

1931. *Wonder Stories,* 2 (February): 1044.

"Letter to the Editor"

1938. *Astronautics,* 40 (April): 17.

1939. *Astronautics,* 44 (November): 10.

Ley, Willy

1932. *Grundriss einer Geschichte der Rakete.* Leipzig: Hachmeister & Thal.

1937. "Rocketeering Or the Hunting of a Canard." *Flight,* 31 (May): 534—535.

1938. *"Raketensensation."* Beilage Nr. 2 in *Astronomische Rundschau,* 1 (April): iii—v.

1961a. "How It All Began." *Space World.* 1 (June): 23—25, 48—50, 52.

1961b. *Rockets, Missiles, and Space Travel.* New York: The Viking Press.

[Lichtenstein, Samuel]

1933. "Treasurer's Report." *Astronautics.* 26 (May): 15.

Limber, Nick

1937a. "What About Rockets?" *Model Airplane News,* 16 (April): 4—5, 30, 32.

1937b. "Rocket Flight In America." *Model Airplane News,* 17 (September): 4—5, 54—56.

1938a. "Double Chamber—The Westchester Rocket Society Will Test New Motor." *Astronautics,* 40 (April): 12.

1938b. "H. Franklin Pierce Discusses Latest Rocket Motors." *American Rocket Society Bulletin,* 1 (September): 1—3.

Locke, George

1975. *Voyages in Space—A Bibliography of Interplanetary Fiction 1801—1914.* Richmond, Surrey (England): Ferret Fantasy Ltd.

Loebell, Ernst

1935. "The Cleveland Rocket Society." *Astronautics,* 32 (October): 13, 18.

Look, Robert

1939. "First Meeting Report." *Journal of the Peoria Rocket Association,* 1 (January): 3. Mimeographed bulletin in collection of Frederick I. Ordway, III.

"The London Branch of the Society Proceedings"

1937. *Journal of the British Interplanetary Society,* 4 (February): 18—19, 13.

"Lt. Cmdr. T. G. W. Settle, The Navy's Ace Balloonist"

1936. *U.S. Air Services,* 21 (May): 18.

"Luftfahrt-Neutigkeiten"

1927. *Der Luftweg,* 15 (August): 204—205.

"Luna"

1969. In Prof. G. V. Petrovich, editor. *The Soviet Encyclopedia of Space Flight:* 230—238. Moscow Mir Publishers.

"Lyon Preparing for Ambitious Rocket Shot"

1931. *Bulletin of the American Interplanetary Society,* 12 (September): 8.

McNash, Caleb

1935. "Are Rocket Motors Practical?" *Popular Aviation,* 16 (March): 150, 182.

1936. "Cooperation Necessary in Rocketry," *Popular Aviation,* 19 (August): 26, 58.

"Mail Sent By Rocket"

1934. *Daily Express,* 18 January 1934.

Malina, Frank J.

1973. "America's First Long-Range Missile and Space Exploration Programme." *Spaceflight,* 15 (December): 442—456.

1974. "On the GALCIT Rocket Research Project, 1936—38." In Frederick C. Durant, III and George S. James, editors, *First Steps Toward Space.* Smithsonian Annals of Flight 10: 113—127. Washington, D.C.: Smithsonian Institution Press.

1977a. "The U.S. Army Air Corps Jet Propulsion Research Project GALCIT Project No. 1, 1939—1946: A Memoir." In R. Cargill Hall, editor, *Essays on the History of Rocketry and Astronautics: Proceedings of the Third Through the Sixth History Symposia of the International Academy of Astronautics:* 153—201. Washington, D.C.: National Aeronautics and Space Administration, NASA Conference Publication 2014.

1977b. "America's First Long-Range Missile and Space Exploration Program: The Ordcit Project of the Jet Propulsion Laboratory, 1943—1946: A Memoir." In R. Cargill Hall, editor, *Essays on the History of Rocketry and Astronautics: Proceedings of the Third Through the Sixth History Symposia of the*

  International Academy of Astronautics: 339—383. Washington, D.C.: National Aeronautics and Space Administration, NASA Conference Publication 2014.

"Manned Rocket Ascends Six Miles!"
  1933. *Space,* mimeographed bulletin of the Cleveland Rocket Society, 1 (December):3.

Marianoff, Dimitri, with Palma Wayne
  1944. *Einstein, An Intimate Study of A Great Man.* Garden City, New York: Doubleday, Doran and Co., Inc.

"Mars Another World"
  1939. *Journal of the Peoria Rocket Society* 1 (January):3.

Martin, Robert E.
  1927. "An Amazing Vision of the Future Scientist Foresees A World Run by Radio." *Popular Science Monthly,* 110 (June):29.

Mayer, D. W. F.
  1937. "A criticism of 'Things to Come.' " *Journal of the British Interplanetary Society,* 4 (February):4—5.

"Membership"
  1934. *Journal of the British Interplanetary Society,* 1 (January):5.

Merkulov, I. A.
  1966. "A Contribution to the History of the Development of Soviet Jet Engineering During the 1930's." In A. A. Blagonravov, et al., editors, *Soviet Rocketry Some Contributions to its History,* NASA TT F-343:41—67. Jerusalem: Israel Program for Scientific Translations.
  1975a. "*Pervenchi Raketnoy Tekhniki.*" *Aviatsiya i Kosmonavtika,* (November):44—45.
  1975b. "*Pervenchi Raketnoy Tekhniki.*" *Aviatsiya i Kosmonavtika.* (December):42—43.

"A Message From President Manning"
  1934. *Astronautics,* 28 (March):1, 8.

"Miscellaneous Rocket News"
  1934. *Astronautics,* 29 (September):8.

"*Midgliederwerbung*"
  1927. *Die Rakete,* (October):138—139.

"*Mitteilungen der Gesellschaft für Weltraumforschung*"
  1938. *Astronomische Rundschau,* 1 (October):supplement No. 3, vii—x.

"*Mitteilungen der Gesellschaft für Weltraumforschung e.V.*"
  1939. *Weltraum,* 1 (December):50—51.

Moreaux, Abbé Th.
  1913. "*Le plus petit objet visible sur la Lune.*" *Cosmos,* 62 (October):282—285.

Morgan, J. H.
  1946. *Assize of Arms—The Disarmament of Germany and Her Rearmament 1919—1939.* New York: Oxford Press.

Moskowitz, Sam
  1954. *The Immortal Storm: A History of Science Fiction Fandom.* Atlanta: The Science Fiction Organization Press.
  1959. *Hugo Gernsback Father of Science Fiction.* New York: Criterios Linotyping & Printing Co., Inc.

Mosely, Leonard
  1974. *The Reich Marshal—A Biography of Herman Goering.* Garden City, New York: Doubleday & Co., Inc.

"*Un Moteur pour aller de la Terre á la Lune.*"
  1913. *Cosmos,* 62 (January):57—58.

"Motor For Rocket Explodes in Test."
  1935. *New York Times,* 26 August 1935.

"*Na Shturm Stratosferi mi gotobi odati svoi sili etomu delu!*"
  1932. *Tekhnika,* 31 (March): page unknown.

[Nebel, Rudolf?]
  1932. "*Raketenflug Bedeutung und Anwendungsmoglichkeiten!*" *Raketenflug Mitteilungsblatt des Raketenflugplatzes Berlin,* 5 (May):1—4.
  1933. "*Die Magdeburger Pilotenrakete!*" *Raketenflug.* 8 (April):1—6.

Nebel, Rudolf
  1932. *Raketenflug.* Berlin: Raketenflugverlag.
  1970. *Raketenflug zum Mond—Von der Idee zur Wirklichkeit.* Dusseldorf: Privately Printed.
  1972. *Die Narren von Tegel.* Dusseldorf: Droste Verlag.
  1977. "Rocket Flight to the Moon—From Idea to Reality." In R. Cargill Hall, editor, *Essays on the History of Rocketry and Astronautics: Proceedings of the Third Through the Sixth History Symposia of the International Academy of Astronautics:* 113—121. Washington, D.C.: National Aeronautics and Space Administration, NASA Conference Publication 2014.

"New Cachet Director for Cleveland Rocket Society"
  1935. *Airpost Journal,* 6 (June):20.

"New Experimental Program"
  1939. *Astronautics,* 43 (August):3.

"New German Society"
  1938. *Bulletin of the Manchester Interplanetary Society.* mimeographed newsletter in Pendray papers, 3 (June):1.

"New Groups Begin Rocket Experimentation"
  1932. *Astronautics,* 23 (October): 4

"News From Abroad"
  1930a. *Bulletin of the American Interplanetary Society.* 1 (June): 3—4.
  1930b. *Bulletin of the American Interplanetary Society.* 2 (July): 4.

"News of the Month—New Groups Begin Rocket Experimentation"
  1932. *Astronautics,* 23 (October): 4.

"News and Views of Current Events"
  1931. *Bulletin of the American Interplanetary Society.* 8 (March—April): 3—6.

"News of the Society"
  1930. *Bulletin of the American Interplanetary Society.* 1 (June): 1.

Norton, Thomas W., and Laurence E. Manning.
  1932. "The Physiology of Acceleration." *Astronautics,* 21 (July): 4—6.

"Notes and News"
  1936. *Journal of the British Interplanetary Society.* 3 (June): 23.
  1937. *Astronautics,* 37 (July): 2, 16.
  1938. *Astronautics,* 39 (January): 3.
  1950a. *Journal of the British Interplanetary Society.* 9 (March): 70—74.
  1950b. *Journal of the British Interplanetary Society.* 9 (May): 137.

"Nouvelles de la Science, Variétés, Informations"
  1938. *Bulletin de la Société Astronomique de France.* 52 (May): 236—237.

Oberg, James E.
  1978. "Korolev and Khushchev and Sputnik." *Spaceflight* 20 (April): 144—150.

Oberth, Hermann
  1923. *Die Raketen zu den Planetenträumen.* Munich: R. Oldenbourg.
  1974. "My contribution to Astronautics." In Frederick C. Durant, III and George S. James, editors, *First Steps Toward Space Flight.* Smithsonian Annals of Flight 10: 129—140. Washington, D.C. Smithsonian Institution Press.

"Oberth Rocket Ready"
  1930. *Bulletin of the American Interplanetary Society.* 4 (October): 8.

O'Donnell, Doris
  1958. "Rocket Pioneer Here Recalls Good Old Days." *Cleveland News,* 28 November 1958.

"Officers of the Society for 1938—39"
  1938. *American Rocket Society Bulletin.* 1 (September): 4.

"Ohioans Plan Rocket Plane"
  1933. United Press story, paper unknown, 7 May 1933, "Ernst Loebell" Biographical File, National Air and Space Museum.

[Ordway, Frederick I, III]
  1978. "On the VfR and *'Die Rakete.' " Spaceflight.* 20 (February): 78—79

Ordway, III, Frederick I., and Mitchell R. Sharpe
  1979. *The Rocket Team.* (New York: Thomas Y. Crowell).

"Organizatsiya v SSSR Obschestva Mezhplanetnykh Soobscheecheniy."
  1924. *Tekhnika i Zhizn,* 12 (July): 1.

"Our Journal"
  1939. *Journal of the Peoria Rocket Association,* 1 (January): 1. Mimeographed bulletin in collection of Frederick I. Ordway, III.

"Our Policy"
  1939. *Journal of the Peoria Rocket Association,* 1 (January): 2. Mimeographed bulletin in collection of Frederick I. Ordway, III.

[Palmer, Ray]
  1967. "Editorial," *Rocket Exchange,* 1 (Summer): 4.

Pastor, Eilert
  1935. *"Der Rückstoss in Nature und Technik,"* Das Neue Fahrzeug, 2 (June): 26—28.

"Paton, Yevgeny Oskarovich"
  1975. In *Bol'shaia Sovetskaia Entsiklopediia.* 19: 280—281. Moscow: Izdatelstvo "Sovetskaia Entsiklopediia."

"M. Pelterie Here"
  1931. *Bulletin of the American Interplanetary Society.* 6 (January): 1.

Pendray, G. Edward
  1931a. "The German Rockets." *Bulletin of the American Interplanetary Society,* 9 (May): 5—12.
  1931b. "Recent Worldwide Advances in Rocketry." *Bulletin of the American Interplanetary Society* 14 (December): 1—8.
  1932a. "The Conquest of Space by Rocket." *Bulletin of the American Interplanetary Society,* 17 (March): 1—7.
  1932b. "The History of the First A.I.S. Rocket." *Astronautics.* 24 (November—December): 1—5.
  1935. "Why Not Shoot Rockets?" *Journal of the British Interplanetary Society.* 2 (October): 9—12.
  1945. *Coming Age of Rocket Power.* New York: Harper & Brothers.
  1955. "The First Quarter Century of the American Rocket Society." *Jet Propulsion,* 25 (November): 586—593.

1963a. "32 Years of ARS History. *Astronautics and Aerospace Engineering,* 1 (February): 124—129.

1963b. "Pioneer Rocket Development in the United States." *Technology and Culture,* 4 (Fall): 384—392.

1974. "Early Rocket Developments of the American Rocket Society." In Frederick C. Durant, III and George S. James, editors, *First Steps Toward Space.* Smithsonian Annals of Flight, 10: 141—151. Washington, D.C.; Smithsonian Institution Press. See also, Adress, Ugo.

"Personliches"

1970. *Astronautik.* 7 (January—February—March—April): 40—42.

"Pervie Astronavti"

1961. *Kosmol'skaya Pravda,* 19 August 1961.

Petrovich, G. V.

n.d. *Development of Rocket Engineering in the USSR.* Moscow?: publisher unknown, no date.

"Petropavlovsky, Boris Sergeyevich"

1969. In G. V. Petrovich, *The Soviet Encyclopedia of Space Flight:* 249. Moscow: Mir Publishers.

Philp, Chas. G.

1937. *Stratosphere and Rocket Flight (Astronautics).* London: Sir Isaac Pitman and Sons, Ltd.

Photograph, Robert Condit's Venus rocket

1928. *Air Travel News,* 2 (April): 19.

Photograph, Tsunendo Obara's rocket

1935. *Astronautics.* 31 (June): 3.

von Pirquet, Guido

1934. *"Zur Frage der Durchführbarkeit der Raumschiffahrt mit den Mitteln der heutigen Technik." Das Neue Fahrzeug,* 3 (June): 22—23.

1939. "Vergleichstabelle diverser Energiewerte." *Weltraum,* 1 (December): 40—48.

"Plans 'Non-Stop Flight' to Venus in Rocket Machine."

1928. *The Washington Star,* 16 February 1928.

"Plans Rocket Ship Flight"

1934. *The Airpost Journal,* 5 (May): 41.

Pobedonostev, Yu. A.

1960. "Behind the Luniks," *Astronautics,* 5 January 1960: 30—50.

1974. "Early Experiments with Ramjet Engines in Flight." In Frederick C. Durant, III and George S. James, editors, *First Steps Toward Space.* Smithsonian Annals of Flight, 10: 167—175, Washington, D.C.; Smithsonian Institution Press.

1974. "First Rocket and Aircraft Flight Tests of Ram Jets." In Frederick C. Durant, III and George S. James, editors, *First Steps Toward Space.* Smithsonian Annals of Flight, 10: 177—184. Washington, D.C.: Smithsonian Institution Press.

Polyarny, A. I.

1974. "On Some Work Done in Rocket Techniques, 1931—38." In Frederick C. Durant, III and George S. James, editors, *First Steps Toward Space.* Smithsonian Annals of Flight, 10: 185—201. Washington, D.C. Smithsonian Institution Press.

"Prämien"

1927. *Die Rakete,* 1 (November): 155.

Pratt, Fletcher

1930. "The Universal Background of Interplanetary Travel." *Bulletin of the American Interplanetary Society,* 1 (June): 2—3.

"Preliminary Rocket Experiments Outlined"

1931. *Bulletin of the American Interplanetary Society.* 12 (September): 4—7.

"President William S. Sykora"

1936. *The International Observer,* mimeographed newsletter of the International Cosmos Science Club, New York, Sam Moskowitz Collection, 1 (November): 4.

Prindle, Jr., Charles A.

1934. "Attaining the Escape Velocity." *Space,* mimeographed bulletin of the Cleveland Rocket Society, 4 (August): 1—3.

"Prof. Dr. A. J. De Bruijn"

1949. *Astro-Jet.* Reaction Research Society, Glendale, California, 24 (Summer): 8—10.

"Profiles—Out of the Ego Chamber"

1969. *The New Yorker Magazine,* 45 (August): 40—42, 44, 46, 51—52, 54—56, 58—65.

"The Proving Stand in Action"

1935. *Astronautics,* 31 (June): 6.

Raushenbakh, B. V., and Yu. V. Biryukov

1974. "S. P. Korolyev and the Development of Soviet Rocket Engineering to 1939." In Frederick C. Durant, III and George S. James, editors, *First Steps Toward Space.* Smithsonian Annals of Flight 10: 203—208. Washington, D.C.; Smithsonian Institution Press.

Razumov, V. V.

1964. "From the History of the Leningrad Group for the Study of Jet Propulsion (LENGSJP)." In USAF Systems Command, Foreign Technology Division, *From the History of Aviation and Cosmonautics (Collection of Articles),* FTD-HT-23-1233-68: 18—34.

"Razumov, Vladimir Vasilyevich"

1969. In Prof. G. V. Petrovich, *The Soviet Encyclopedia of Space Flight:* 333. Moscow: Mir Publishers.

"The Reader Speaks—Interplanetary Society Progress"
  1930. *Wonder Stories,* 2 (December):754, 757—765.
"The Reader Speaks"
  1931a. *Wonder Stories,* 2 (February):1044—1053.
  1931b. *Wonder Stories,* 2 (March):1195—1196.
  1931c. *Wonder Stories,* 2 (April):1332—1342.
"Report of Motor Tests of June 2nd"
  1935. *Astronautics,* 32 (October):3—4, 20.
"Report of the President of the American Interplanetary Society"
  1932. *Astronautics,* 19 (May):7.
"The Report of Rocket No. 3"
  1934. *Astronautics,* 30 (October—November):5.
"Report on the 1938 Rocket Motor Tests"
  1939. *Astronautics,* 42 (February):1—5.
"Report of Motor Tests of June 8, 1941
  1941. *Astronautics,* 49 (August):3—5.
Riabchikov, Evgeny
  1971. *Russians in Space.* New York: Doubleday & Co.
Rinin [sic], N. A.
  1935. *"Propulsione a reazione senza utilizzazione del'aria esterra."* L'Aerotecnica, 15 (September—
         October):912—914.
"Rocket Artillery Latest Menace to Civilization"
  1934. *The Japan Times & Mail,* 6 October 1934.
"Rocket Demonstration at New York World's Fair"
  1939. *Astronautics,* 43 (August):16.
"Rocket Experiments of 1934"
  1934. *Astronautics,* 29 (September):1—3, 9.
"Rocket Motor Tests of October 20, 1935"
  1936. *Astronautics,* 34 (June): 5, 13, 20.
"Rocket Society Affiliates Two Groups Join Under New Decentralizing Plan"
  1937. *Astronautics,* 37 (July):12.
"Rocket to Spread Molten Metal"
  1924. *The Times,* London, 11 June 1924.
"Rocket to Make War Horrible"
  1931. *Bulletin of the American Interplanetary Society* 7 (February):8.
Ross, H. E.
  1939. "B.I.S. Space-Ship." *Journal of the British Interplanetary Society,* 5 (January):4—9.
  1950. "Gone With the Efflux." *Journal of the British Interplanetary Society,* 9 (May):93—101.
  1969. "The B.I.S. Spaceship." *Spaceflight,* 11 (February):42—47.
  1974. "The British Interplanetary Society's Astronautical Studies, 1937—39." In Frederick C. Durant, III
         and George S. James, editors, *First Steps Toward Space.* Smithsonian Annals of Flight 10:209—
         216. Washington, D.C.: Smithsonian Institution Press.
"Rumor Being Traced by CRS"
  1933. *Space,* mimeographed bulletin of the Cleveland Rocket Society, 1 (December):3.
Rynin, N. A.
  1935. See also Rinin, N. A.
  1970— *Interplanetary Flight and Communication (Mezhplanetnye soobshcheniya),* NASA TT F-640 to
  1971 NASA TT F-648, 9 volumes. Jerusalem: Israeli Program for Sicentific Translations.
Sänger, Eugen
  1936. "The Rocket Combustion Motor." *Astronautics,* 3 (October): 2—12. Translation by Merritt A.
         Williamson.
Sänger-Bredt, Irene, and Rolf Engel
  1974. "The Development of Regeneratively Cooled Liquid Rocket Engines in Austria and Germany,
         1926—42." In Frederick C. Durant, III and George S. James, editors, *First Steps Toward Space.*
         Smithsonian Institution Press.
"Satzungen der Gesselschaft für Weltraumforschung"
  1939. *Weltraum,* 1 (April):30—32.
"Schachner Ascends to Society's Presidency"
  1932. *Astronautics,* 22 (August—September):2—3.
Scherschevsky, A. B.
  1929.*Die Rakete für Fahrt und Flug.* lin: C. J. E. Volckmann.
"The Scienceers"
  1930.*Wonder Stories,* 1 (Ma):1139.
"Seek to Better Rocket Motor"
  1934. *Cleveland News,* 25 June 1934.
Shchetinkov, Yevgeny S.
  1974. "Development of Winged Rockets in the USSR, 1930—39." In Frederick C. Durant, III and George
         S. James, editors, *First Steps Toward Space.* Smithsonian Annals of Flight 10:247—257. Wash-
         ington, D.C.: Smithsonian Institution Presss.

1977.   "Main Lines of Scientific and Technical Research at the Jet Propulsion Research Institute (RNII), 1933—1942." In R. Cargill Hall, editor, *Essays On the History of Rocketry and Astronautics: Proceedings of the Third Through the Sixth History Symposia of the International Academy of Astronautics*, NASA Conference Publication 2014, 2(43—57). Washington, D.C.: National Aeronautics and Space Administration.

Shesta, John
1935.   "Report on Rocket Tests." *Astronautics*, 31 (June): 4—5.

Shesta, John; H. Franklin Pierce; and James H. Wyld.
1939.   "Report on the 1938 Rocket Motor Tests." *Astronautics*, 42 (February): 1—5.

Shesta, J., and R. Healey
1941.   "Report on Motor Tests of June 8, 1941 at Midvale." *Astronautics*, 49 (August): 3—5.

Shlykova, S. A.
1966.   "K. E. Tsiolkovskii's Correspondence with the Jet Scientific Research Institute." In A. A. Blagonravov, et al, editors, *Soviet Rocketry Some Contributions to its History*, NASA TT F-343: 127—132. Jerusalem: Israel Program for Scientific Translations.

Shutto, I.
1973.   " 'Katyusha' i Drugii." *Znamya*, (8 August): 173—195.

"A Skyrocket Flying Machine"
1913.   *Scientific American*, 108 (March): 289.

Skoog, A. Ingemar
1974.   "Wilhelm Theodore Unge: An Evaluation of His Contributions." In Frederick C. Durant, III and George S. James, editors, *First Steps Toward Space*. Smithsonian Annals of Flight 10: 259—267. Washington, D.C.: Smithsonian Institution Press.

Slukhai, I. A.
1968.   *Russian Rocketry A Historical Survey (Rakety i Traditsii)*. Jerusalem: Israel Program for Scientific Translations. NASA TT F-426.

Smith, Bernard
1940.   "Liquid Cooling For Rocket Motors." *Astronautics*, 47 (November): 10—11, 15.

Smith, R. A.
1942.   "Lunar Space Vessel." *Flight*, 41 (February): d—f.

Smith, R. A. , et al.
1937.   "Technicalities." *Journal of the British Interplanetary Society*, 4 (December): 8—15.

Smits, Rudolf, Compiler
1968.   *Half A Century of Soviet Serials 1917—1968*. Washington, D.C.; Library of Congress.

"Society Elects 1932—1933 Officers"
1932.   *Astronautics*, 19 (November): 1.

"Society Holds First Annual Meeting"
1931.   *Bulletin of the American Interplanetary Society*. 8 (March—April): 1—3.

"Society Incorporates"
1931.   *Bulletin of the American Interplanetary Society*. 9 (May): 1.

"Society Offered Free Propane"
1934.   *Space*, mimeographed bulletin of the Cleveland Rocket Society, 2 (January—February): 6.

"Society's Rocket To Be Exhibited"
1932.   *Bulletin of the American Interplanetary Society*, 16 (February): 6.

"Society's Rocket To Meet Final Tests."
1932.   *Bulletin of the American Interplanetary Society*, 17 (March): 7—8.

"Society for Studying Interplanetary Travel"
1969.   "In Prof. G. V. Petrovich, editor, *The Soviet Encyclopedia of Space Flight*: 377. Moscow: Mir Publishers.

Sokol'skii, V. S.
1967.   *A Short Outline of the Development of Rocket Research in the USSR*. NASA TT F 67-51340. Jerusalem: Israeli Program for Scientific Translations.

"Soviet Engineers Constructing Two Rockets"
1932.   *The American Interplanetary Society Bulletin*. 15 (January): 1.

"Soviet Orders A Rocket That Will Rise 34 Miles"
1935.   *New York Times*, 12 May 1935.

"Space Flying"
1930.   *The Planet*, mimeographed magazine of the Scienceers, New York, Sam Moskowitz Collection, 1 (September): 1.

Spangenberg, Karl
1934.   "Ultra-short Wave Antenna for Plotting Rocket Trajectories." *Space*, mimeographed bulletin of the Cleveland Rocket Society, 4 (August): 4—9, 12.

Steinhoff, Ernst A.
1974.   "Early Developments in Rocket and Spacecraft Performance, Guidance, and Instrumentation." In Frederick C. Durant, III and George S. James, editors, *First Steps Toward Space*. Smithsonian Annals of Flight, 10: 277—285. Washington, D.C.; Smithsonian Institution Press.

Steinitz, Otto
1933.   "*Von der Berliner Auto-Ausstellung.*" *Das Neue Fahrzeug*, 1 (February): 2—3.
1937.   "*Zur Stabilität von Weltraum-Raketen.*" *Das Neue Fahrzeug*, 12 (May): 1—2.

139

Sternfeld, Ari

    1959. *Soviet Space Science.* New York: Basic Books, Inc.

Stoiko, Michael

    1970. *Soviet Rocketry: Past, Present, and Future.* New York: Holt, Rinehart and Winston

Stranger, Ralph

    1936. "A Three-Year Old Mystery." *Journal of the British Interplanetary Society,* 3 (June):17—18, 21.

"."'Strato' Rocket Planned in Soviet"

    1935. *New York American,* 9 May 1935.

"Stratosphere Rocket"

    1935. *New York Times,* 14 July 1935.

Strong, J. G.

    1934. "Interplanetary Societies—Are They Too Fictitious?" *Journal of the British Interplanetary Society.* 1 (October):33.

"Studying Rocket Ships Is His Hobby"

    1936. *Peoria Star* (Peoria, Illinois) 18 March 1936.

"Support"

    1939. *Journal of the Peoria Rocket Association,* 1 (January):4. Mimeographed bulletin in collection of Frederick I. Ordway, III.

Sykora, Fritz

    1960. "Pioniere der Raketentechnik aus Österreich." In *Technisches Museum für Industrie und Gewerbe in Wien Forschungsinstitut für Technikgeschichte Blätter für Technikgeschichte,* 22:189—204. Vienna: Springer-Verlag.

Sykora, William S.

    1935. "A Message." *The International Observer,* mimeographed newsletter of the International Cosmos Science Club; New York, Sam Moskowitz Collection, 1 (November):1.

    1936. "A Miniature Rocket Motor." *The International Observer,* mimeographed newsletter of the International Cosmos Science Club, Sam Moskowitz Collection, 1 (March):8—11.

Tascher, John

    1966. "U.S. Rocket Society Number Two, The Story of the Cleveland Rocket Society." *Technology and Culture,* 7 (Winter):48—63.

"Test Report on Rocket No. 3."

    1934. *Astronautics,* 30 (October—November):5—6, 11.

Thomas, Shirley

    1960—

    1968. *Men of Space,* Philadelphia: Chilton Co. (8 volumes).

"Tikhomirov, Nikolai"

    1969. In G. V. Petrovich, editor, *The Soviet Encyclopedia of Space Flight:* 462. Moscow: Mir Publishers.

Tikhonravov, M. K.

    1965. *K. E. Tsiolkovsky Works on Rocket Technology,* NASA TT F-243. Washington, D.C.: National Aeronautics and Space Administration.

    1974. "From the History of Early Soviet Liquid-Propellant Rockets." In Frederick C. Durant, III and George S. James, editors, *First Steps Toward Space,* Smithsonian Annals of Flight 10:287—293.

Tikhonravov, J. K., and V. P. Zaytsev

    1977. "On The History of the Stratosphere Rocket Sonde In the USSR, 1933—1946." In R. Cargill Hall, editor, *Essays on the History of Rocketry and Astronautics: Proceedings of the Third Through the Sixth History Symposia of the International Academy of Astronautics,* 2:65—78. Washington, D.C.: National Aeronautics and Space Administration.

Tokaty, G. A.

    1968. "Foundations of Soviet Cosmonautics," *Spaceflight,* 10 (October): 335—346.

"To Publish German Translations:

    1931. *Bulletin of the American Interplanetary Society,* 6 (January):8.

"Two Firing Tests Made"

    1933. *Space,* mimeographed bulletin of the Cleveland Rocket Society, 1 (December):9.

"Two Thousand At Museum Meeting"

    1931. *Bulletin of the American Interplanetary Society,* 8 (February):1.

"Traveling Through Inter-stellar Space What Type of Motor Would You Employ?"

    1913. *The Scientific American Supplement.* 75 (April): 263.

"Treasurer's Report"

    1933. *Astronautics,* 26 (May):15.

"Trophy"

    1933. *Space,* mimeographed bulletin of the Cleveland Rocket Society, 1 (December):3.

Tsander, F. A.

    1969. *From A Scientific Heritage,* NASA TT F-541. Washington, D.C.: National Aeronautics and Space Administration.

Tsirkov, B.

    1968. "Pilot, Uchenbia, Populurizator." *Aviatsia i. Kosmonavtika,* (June):22—26.

Uhler, Harry B.

    1978. "Semicentennial: Baltimore-To-Venus Attempt." *Science News,* 114 (July):78—79.

"University News—Lectures"

    1938. *Yale Alumni Magazine,* 1 (February): 7.

U.S. Army Technical Intelligence Center, Tokyo

    1946. *Ordnance Technical Intelligence Report Number 4—Subject: "Survey of Japanese Rocket Research and Development."* Prepared by Captain Joseph R. Weeks, Tokyo: U.S. Army Technical Intelligence Center.

Van Dresser, Peter

    1936. "The Rocket Motor A Brief Survey of Ideas Concerning its Design and Construction," *Astronautics,* 33 (March): 9— 13, 18.

"Verein für Raumschiffahrt E. V."

    1927. *Die Rakete,* 1 (July): 82—84.

Versins, Pierre

    1972. *Encyclopedie de L'Utopie.* Lausanne: *L'Age d'Homme.*

Vocke, Susanne

    1938. *"Die Offentlichen Vortrage des Vierteljahres Oktober bis Dezember 1937." Astronomische Rundschau,* 1 (January): 15— 17.

'Volanzan' (Cohete Volador) Organo Oficial de Divulgación del Centro de Estudios Astronáuticos 'Volanzan'

    1932. Buenos Aires (Argentina), 1 (December): 1—4. Leaflet-journal in the collection of Herbert Schaefer.

Waterhouse, Helen S.

    1934. "Has the Rocket A Future?" *Popular Aviation,* (November: 301, 335.

Wellman, Bertha

    1944. "Pioneered in Stratosphere Rocket Mottos [sic] Tests." *The Cleveland Press,* 8 December 1944.

Williams, Beryl, and Samuel Epstein

    1962. *The Rocket Pioneers on the Road to Space.* New York: Julian Messner.

Williamson, Merritt

    1965. "The Yale Rocket Club 1935— 1940." *Yale Scientific Magazine,* 39 (March): 16— 18.

"Willy Ley, 62, Prolific writer on Scientific Subjects, Is Dead"

    1969. *New York Times,* 25 June 1969.

[Winkler, Johannes]

    1928. *"Rückstosse-Diagram einer Feuerwerksrakete." Die Rakete,* 2 (January): 3— 5.

Winter, Frank H.

    1977. "Birth of the VfR: The Start of Modern Astronautics." *Spaceflight,* 19 (August): 243— 256.

"Wire-Tailed 'Snare' Rockets Defend Small Ships from Air Attacks"

    1943. *Astronautics,* 54 (February): 8—9.

"Woman Injured in Rocket Blast"

    1935. *New York Times,* 22 October 1935.

Wyld, James H.

    1934. 'The Problem of Rocket Fuel Feed." *Astronautics,* 34 (June): 8— 13.

    1936. "Fundamental Equations of Rocket Motion." *Astronautics,* 35 (October): 13— 16.

    1938. "Fuel as Coolant." *Astronautics,* 40 (April): 11— 12.

"Wyld Motor Retested"

    1941. *Astronautics,* 50 (October): 8.

"Yale Clubs Stress Study of Air Use"

    1937. *New York Times,* 2 May 1937.

"Yale Group Studies Rockets"

    1937. *Modern Mechanix,* 18 (September): 41.

"Young 'Rocketeers' Will Go On—Five Injured By Exploring Rocket at Clayton"

    1937. *The Reporter* (Ashton Under Lyne, Lancashire, England), 2 April 1937.

Notes

1. Interview, Africano, 1976; interview, Loebell, 1978.
2. Von Braun and Ordway, III, 1969:9—13.
3. Freedman, 1965:34, 125.
4. Mendillo and DeVorkin, 1977:9; Versins, 1972:335—337. "Good Day," 1919:525.
5. Locke, 1975:67—69.
6. Lehman, 1963:22—23; Blossett, 1974:8—9.
7. "Skyrocket," 1913:289; "Travelling," 1913:263; "Moteur," 1913:57—58; "Autres," 1913:277; Moreaux, 1913:282—385.
8. Lehman, 1953:103—104.
9. Oberth, 1923:24; Goddard, 1919:56—57; Goddard, Esther and Pendray, 1970:514.
10. Ley, 1961b:112—113, 508.
11. Tikhonravov, 1965:11,85; Brügel, 1933:114—118.
12. Winter, 1977:244; Oberth, 1974:136.
13. Korneev, 1964:3,9,15—16.
14. "Pervie," 1951:4.
15. "Pervie," 1961:20—25
16. Rynin, 1971:1(3)46; Rynin, 1971:3(9)168—170.
17. Essers, 1976:18—19, 253—254; Ley, 1961b:116,530,532.
18. Ley, 1961b:91—100.
19. Aldiss, 1973:210; Kyle, 1976:135.
20. Heflin, 1969:46,47; "REP-Hirsch Prize," (Misc.).
21. Giles, 1944:12—13.
22. "Society," 1969:377; Korneev, 1964:18—19; Stoiko, 1970:31; Sokol'skii, 1967:12; Tokaty, 1963:335; "Pervie," 1961:4.
23. "Organizatsiya," 1924:1; Smits, 1968:(2)990.
24. "Society," 1959:377; Merkulov, 1966:43; Riabchikov, 1971:120.
25. Korneev, 1964:19.
26. Merkulov, 1966:43; Korneev, 1964:19; Lehman, 1963:110—111.
27. Korneev, 1964:19.
28. Rauschenbakh and Biryukov, 1974:203.
29. Glushko, 1973:11; "Grave," 1972:(7)197; "Paton," 1975:(19)280; "Pervie," 1961:4.
30. Rynin, 1971:1(1)19—20.
31. "Welsh," (Misc.); "Death," 1924:40; "Death Dealing," 1924:656; "Rocket," 1924:9; Martin, 1927:29; Rynin, 1971:2(4)166.
32. Rynin, 1971:2(4)201; Rynin, 1971:1(3)30—31.
33. Rynin, 1971:2(4)205—206; Essers, 1976:135; "Luna," 1969:231—232.
34. Sykora, 1960:195—196; Essers, 1976:238; Rynin, 1971:1(1)21.
35. Sykora, 1960:196—197; "Ing. Guido," 1950:205—206; "Guido von Pirquet," 1966:23; Sykora, 1980.
36. Essers, 1976:61—64; Sykora, 1980.
37. Essers, 1976:90—91.
38. Essers, 1976:94.
39. Essers, 1976:119—120.
40. Essers, 1976:121; Rynin, 1971:1(1)20—21.
41. Essers, 1976:123.
42. Essers, 1976:124.
43. Ley, 1961b:440; Sänger-Bredt and Engel, 1974:229—230; Letter, Sänger-Bredt, 1977; "Esposizione Aeronautica," 1928:308, Brügel, 1933:71,62.

    Von Pirquet also spoke, presumably representing the Austrian Society for High Altitude Exploration, before American students at the Austro-American Society at Vienna University on 8 July 1928. Von Pirquet also corresponded with the leading—or rising—members of the VfR, The German Rocket Society, namely: Wernher von Braun, Willy Ley, Hermann Oberth, and Rudolf Nebel.
44. Ley, 1961b:440—443; "Austria," 1934:17; Rynin, 1971:2(4)91—95; Winter, 1977:248; Letter, Ley-Pendray, 2 November 1931; Brügel, 1933:64.
45. Haley, 1958:282.
46. "Verein," 1927:82—84; "Luftfahrt-Neutigkeiten," 1927:205.
47. Ley, 1961a:21—25.
48. Essers, 1976:129, 134.
49. Rynin, 1971:1(4)201; Essers, 1976:136.
50. Essers, 1976:137—138; Ley, 1961b:117.
51. Ley, 1961b:117; "Verein," 1927:82—84.
52. "Verein," 1927:82—84.
53. "Verein," 1927:82—84; Ley, 1961b:passim; Nebel, 1972:96.
54. "Mitgliederwerbung," 1927:138—1939; "Ein Jahr," 1928:99—100; Ley, 1961b:118, 131—136; Winter, 1977:247; "Ein Brief," 1929:43—44.
55. "Mitgliederwerbung," 1927:138—139; "Prämien," 1928:151; "5000 RM," 1927:155; Palmer, 1967:4; Letter, Browne, 1978; Letter, Palmer, 1977.

    Early VfR support may have also come from America science fiction enthusiast and later publisher

Ray Palmer, now deceased, claimed to have raised $2,000 in 1927 through the Milwaukee Fictioneers science fiction club. Long time associate of Palmer, Howard Browne, doubts the story as Palmer was 16 in 1927. Palmer's widow had no recollection of her husband telling the story.

56. Letter, VfR, 1931; interview, von Khuon, 1977; Ley, 1961b:151; Letter, Ley, 1931a; Letter, Ley, 1931b; "CRS," 1933:2.

So far as is known, the VfR and the Cleveland Rocket Society were the only major astronautical societies of the 1930s that had pins. G. Edward Pendray and P. E. Cleator say that neither of their groups had any. Lapel buttons were discussed in the British Interplanetary Society meeting of 3 August 1934, but no action was taken.

57. Ley, 1961b:118; Essers, 1976:138; "Bekanntmachung," 1927:142; "Amtliche," 1928:34; Ordway III, 1978:78—79; Scherschevsky, 1929:90.

58. Winkler, 1928:3—5; Ley, 1961b:145; Sänger-Bredt and Engel, 1974:219, 221.

59. Essers, 1976:207, 209—210; Letter, Ley, 1934; Winter, 1977:248, Cleator, 1934:13.

60. Ley, 1961b:126—127.

61. Nebel, 1972:36—42, 44, 55, 63; Nebel, 1977:113—121; Mosley, 1974:24—27.

Unsaid by Nebel is that the French began using their air-to-air Le Prieur rockets in 1916—in the Somme, where Nebel was stationed. These rockets were very successful against German Zeppelins and captive observation balloons. Nebel's own unit, Jagdstaffel 5, was equipped with such balloons, besides planes. It should be noted that Hermann Oberth also proposed a plan for rockets during the war. In 1917 he suggested a liquid air-alcohol rocket to the German War Department but was turned down.

62. Nebel, 1970:4—6.

63. Gartman, 1956:64—65; Ley, 1961b:127; Oberth, 1974:139—140.

64. Ley, 1961b:124—131; "Errichtung," 1929:99; Ley, 1931:15, 349—350; Winter, 1977:249.

65. Ley, 1961b:132—134; Sänger-Bredt and Engel, 1974:220.

66. Schultz, 1979:11; Ley, 1961b:153; Brügel, 1933:15—20; Von Dickhuth-Harrach, 1928:6.

67. Ley, 1961b:135; Nebel, 1970:7.

68. Ley, 1961b:136—8; Nebel, 1972:86—87; Gartman, 1956:84; 1977.

69. Ley, 1961b:137—8; Bergaust, 1976:41; Nebel, 1970:9; Von Braun, 1967:35—36; Nebel, 1972:96—97; Interview, Schaefer, 1977.

70. Interview, Schaefer, 1977; Nebel, 1972:94—95; Interview, Engel, 1975.

71. Nebel, 1972:74, 114, 121—123; Gartmann, 1956:85—6; Marianoff and Wayne, 1944:115—119; Blosett, 1974:10—11; Pendray, 1974:141—142; Drummond-Hay, 1930:24—26; Cleator, 1934:13; Ley, 1961b:145; Interview, Delaney, 1979.

Once under way the Raketenflugplatz received many visitors, some of them very distinguished. In September 1930, hopes were entertained that Henry Ford would come when he toured Germany. Nebel sent him a telegram offering him a liquid-fuel rocket for the Ford Museum in Dearborn, Michigan, in exchange for VfR patronage. Ford never responded. However, the French banker and cofounder of the REP-Hirsch Astronautical Prize, Hirsch, came in the Spring of 1931. American Interplanetary Society President G. Edward Pendray visited in April of that year. Other visitors were Phillip E. Cleator, founder of the British Interplanetary Society; Dr. Franz Hermann Ritter of the Chemisch-Technische Reichsanstalt who had earlier certified Oberth's Kegelduse motor; Vladimír Mandl, the future space law pioneer; Reinhold Tiling, the German private rocket experimenter; and the German Minister of the Air Sport Union, Dominicus. Newsmen were also frequent visitors, some of them not always positive. British aviatrix—newswoman Lady Drummond-Hay came in 1930.

72. Nebel, 1982:95—96.

73. Marianoff and Wayne, 1944:115.

74. Ley, 1961b:151, 157; Interview, Schaefer, 1977; Nebel, 1932:1—4.

75. Ley, 1966:147—152; Letter, Ley, 1931a.

76. Letter, Ley, 1931c.

77. Ley, 1961b:151—2; Nebel, 1972:117—119; "Verein," (Misc.); Letter, Ley, 1931b.

78. Ley, 1951b:152, 161; Interview, Schaefer, 1977; "Verein," (Misc.).

79. Ley, 1961b:157—158; Nebel, 1972:125—126.

80. Ley, 1961b:157—158; Nebel, 1972:125—126; "Verein" (Misc.); Nebel, 1933:1—6.

81. Letter, Ley, 1933; Interview, Schaefer, 1977; Nebel, 1972; Hüter, (Misc.), 1945:1—2; Interview, Schaefer, 1977; Sänger-Bredt and Engel, 1974:226—227; Von Dickhuth-Harrach and Letter, Ley, 1.

82. Interview, Schaefer, 1977; Sänger-Bredt and Engel, 1974:226—227; Schaefer, (Misc.), 1932—1936; Nebel, 1933:1—6; Ley, 1961b:158—160; Nebel, 1972:118—121; 127—129.

83. Ley, 1961b:160; Sänger-Bredt and Engel, 1974:226; Interview, Schaefer, 1977; Schaefer (Misc.), 1932—1936.

84. Sänger-Bredt and Engel, 1974:passim; Interview, Schaefer, 1977; Schaefer (Misc.), 1932—1936.

85. Ley, 1961b:157, 161; Letter, Ley, 1933; Von Dickhuth-Harrach and Letter, Ley, 1934; Nebel, 1972:123—124; Ley, 1961b:161; Interview, Schaefer, 1977.

86. Letter, Ley, 1934; Ley, 1961b:199; (Letter) Schlesinger, 1955; Giles, 1944:12—13; Interview, Ehricke, 1978; Von Pirquet, 1934:22—23; Steinitz, 1937:1—2; Von Dickhuth-Harrach, 1937:2—4; "Friedrich," 1968:131; Letter, Cleator, 1934d; Letter, Cleator, 1935.

87. Schershevsky, 1929:90; Sänger-Bredt and Engel, 1974:237—240; "Albert Püllenberg" (Misc.); Dornberger 1954:16—17, 248; Kaiser, 1954:49—51, 70; Interview, Ehricke, 1978; Koelle, 1955:121—122; Kaiser, Letter, 1939; "New German," 1938:1; Ley, 1938:iii—v; "Mitteilungen," 1939:vii—x; "Mitteilungen," 1939:50—51.

Hans K. Kaiser did later write his book on rockets. In fact, there have been several. For Arthur C. Clarke's review of one of Kaiser's works, *Kleine Raketenkunde,* see *Journal of the British Interplanetary Society,* 9 (July 1950):206.

The formerly secret *German Organisation and Personalities Engaged in Research and Development of Armaments During the Second World War,* prepared in 1948 by the Technical Information Bureau of the British Ministry of Supply, shows that Hermann Oberth was in Kaiser's group and even subordinated to him. Kaiser is given as a mathematician-physicist. Oberth was in charge of Planning in the Study Group for New Ideas.

88. Cranz, 1925:403—419; Ley, 1961b:155, 198; Thomas, 1951:(2)48; Dornberger, 1958:28; Dornberger, 1954:248—249; Hartl, 1958:157—168; Helfers, 1954; 47—48; Morgan, 1946:*passim;* Bullock, 1952:314.

89. Dornberger, 1958:27, 31—32; Interview, Schaefer, 1977; Ley, 1961b:155—145; Ordway III and Sharpe, 1979:18.

On 3 August 1936 Riedel and Nebel were granted German patent No. 633,667 for *Rückstossmotor für flüssige Triebstoffe* (Reaction Motor for Liquid Propellant). Von Braun, in order to gain a valuable technician for the V-2 project at Peenemünde, concocted a "secret contract," dated 2 July 1937, whereby the Army purchased the patent for 75,000 Marks (50,000 for Nebel and 25,000 for Riedel— with "voluntary gratifications" of 5,000 Marks each to Heinisch, Hüter, Bermüller, and Ehmeyer). The payment was "due on the day Riedel reports for service with the *Waffenprufamt* [Weapon Proofing Office, actually Peenemünde]." According to von Braun many years later, in a letter written to Willy Ley: "I was definitely not interested in Nebel, whom I had always considered a successful if unscrupulous salesman with little technical and no scientific background. Klaus Riedel, who fully shared my view that Nebel would never fit into the Peenemünde framework, felt that it would put him into an embarrassing position if he were to take service with Peenemünde and have to watch Nebel go empty-handed. While Klaus too considered Nebel a 'dud' in technical matters, he pointed out that Nebel, through his tireless salesmanship, had given him the opportunity to work in liquid-fuel rocket development and thereby to acquire the experience which made him valuable for Peenemünde . . . Dornberger and I finally hit upon the idea of purchasing the old Nebel-Riedel patent . . . This system had been abandoned at Kummersdorf as early as 1934 and there was, consequently, neither a technical nor a legal reason for the *Heereswaffenamt* to acquire the patent. The purchase was nothing but a convenient bureaucratic vehicle to pay Nebel off."

90. Dornberger, 1958:xvii, 16—17, 26—27; Interview, Schaefer, 1977; Bergaust, 1976:47—8; Thomas, 1950:(1)137; Von Braun, 1950:87—88; Letter, Zoike, 1978; Ley, 1961b:450; Nebel, 1972:133—151; Ordway III and Sharpe, 1979:19, Von Braun, 1967:39.

91. Petrovich, no date:13; Sokol'skii, 1967:13; Rynin, 1971:3(9)3, 177; Rynin, 1971:3(7)31; Haley, 1958:230.

92. Glushko, 1974:30—31; *"Tikhomirov,"* 1969:462; Goddard, Esther, and Pendray, 1970:18; Skoog, 1974:260—261.

93. Ritchie, (Misc.):2—4; *"Petrovavlovsky,"* 1969:299; Kulagin, 1974:91; *"Glushko,"* 1968:181.

94. Glushko, 1975:1; Glushko, 1974:32—33; Stoiko, 1970:370; Goddard, Esther and Pendray, 1970:13, 22, 246.

95. Golovanov, 1975:212; Petrovich, no date:20—21; Glushko, 1973:14—15; Tokaty, 1968:337; Rauschenbakh and Biryukov, 1974:204; Lee, 1962:29—30; Korneev, 1964:38—39; Tikhonravov, 1972:288. 96. *"Na Shturm,"* 1932:page unknown.

97. Korneev, 1964:32—37; Tsander, 1969:52.

98. Glushko, 1973:10; Korneev, 1964:36—38; Tsander, 1969:86—87.

99. *"Korolev,"* 1972:465; Blagonravov, 1968:9; Golovanov, 1975:169—170, 201—208.

100. Golovanov, 1975:217—218.

101. Golovanov, 1975:218—220; Kulagin, 1984:93.

102. Golovanov, 1975:234—235.

103. Korneev, 1964:46—55; Rauschenbakh and Biryukov, 1974:203—204.

104. Golovanov, 1975:236, 241, 244, 250—252, 274; Rauschenbakh and Biryukov, 1974:204; Tikhonravov, 1974:288; Shchetinkov, 1974:248.

105. Golovanov, 1975:247—248.

106. Golovanov, 1975:248—249; Rauschenbakh and Biryukov, 1974:205.

107. Glushko, 1973:15—16; Golovanov, 1975:265—268, 275—277; *"Kleimenov,"* 1969:209 Tokaty, 1968:339.

108. Razumov, 1964:18—20; *"Razumov,"* 1969:333.

109. Razumov, 1964:20; Glushko, 1973:12; Letter, Cleator, 1936a; Letter, Ley, 1931c.

110. Golovanov, 1975:278—279; Razumov, 1964:18—34; *"Leningrad,"* 1969:217—218.

111. Rynin, 1971:3(9)2—4; *"Rynin,"* (Misc.); Brügel, 1933:137—140; Rinin (sic.), 1938:912—914; Letter, Cleator, 1936a; Letter, Ley, 1931c; Gartman, 1956:96; "Soviet Orders," 1935; "Strato," 1935; "Stratosphere," 1935.

Cleator to Pendray, 10 March 1936: "You ask about the USSR experiments. I must confess that I have practically no information. I seem, indeed, to be in exactly the same position as yourself. Repeatedly I have requested details, and repeatedly I have received a nice acknowledgement of my letter, and precious little else!" Some distorted stories of Soviet rockets, however, appeared in Western papers in the 30's. See last three citations as examples.

112. Golovanov, 1935:279—280; Shchetinkov, 1974:248; Glushko, 1973:17—20.

113. Sklykova, 1966:127—128; Golovanov, 1975:257—258.
114. Shlykova, 1966:129—132; Gartmann, 1956:34.
115. Glushko, 1973:19.
116. Merkulov, 1966:57; Hingley, 1974:257, 262; Glushko, 1975:21—22, 35; Glushko, 1973:15; Tokaty, 1968:341; Slukhai, 1968:39; Oberth, 1978:144—150; Shutko, 1973:176, 183—185.

Tukachevsky's role in Soviet (military) rocketry is brought out in more detail by I. Shutko in his article "'Katyusha' i Drugi," in Znamya 8 (August 1973); (1973):176, 183—185. The GIRD X Launch picture in this book shows 11 people. Another version of the picture shows another man on the right.
117. Shchetinkov, 1977:43—63; Merkulov, 1975a:45; Merkulov, 1975b:43.
118. Tikhonravov, 1974:287—290; Riabchikov, 1971:109—113.
119. Riabchikov, 1971:114; Shchetinkov, 1974:248—257.
120. Langmak and Glushko, 1935:93—94.
121. Astashenkov, 1971:86, 91, 76—79.
122. Raushenbakh and Birynkov, 1973:204—207; Shchetinkov, 1977:(2)51—53; Shchetinkov, 1974:248—254; Astashenkov, 1971:95.
123. Polyarny, 1984:185—186; Tikhonravov and Zaytsev, 1977:65—68; Razumov, 1964:23—24, 27—34.

The use of cameras in sounding and other rockets did not originate with the Soviets. Goddard installed a camera in his 1929 liquid-fuel rocket flight, but the rocket crashed. The earliest history of "camera rockets" is found in the paper, "Camera Rockets and Space Photography Concepts Before World War II," by Frank H. Winter, read at the Seventh International History of Astronautics Symposium, October 1973, Baku, USSR.
124. Tikhonravov and Zaytsev, 1977:69—70; Polyarny, 1974:187—191, 194—198.
125. Tikhonravov and Zaytsev, 1977:74—75; Polyarny, 1974:198—200; Glushko, 1973:19.
126. Tikhonravov, 1974:291; Tikhonravov and Zaytsev, 1977:70—72; Merkulov, Dushkin, 1977:81.
127. Tikhonravov and Zaytsev, 1977:72—74; Pobedonostev, 1977:167—175.

For a good overall account of Soviet ramjet developments in the 1930s and 40s, see Richard P. Hallion, "The Soviet Stovepipes," Air Enthusiast, 9, February—May 1979:55—60.
128. Speech, Tokaty, 1968; Pobedonostev, 1960:50; Ordway III and Sharpe, 1979:342; Glushko, 1977:228—229, 460—493.

V. P. Glushko's massive (503 pages) Putyi v Raketnoi Tekhniki Izbranni Trudi 1924—1946 (Road to Rocket Technology—Selected Work 1924—1946) is clearly the most complete account of Soviet technical accomplishments in rocketry during the earliest years. Unfortunately, however, the true "inside" story of the various Soviet rocketry and astronautical societies is still lacking from the Soviet side. Glushko devotes only the last few pages (450—491) to his personal history and to general organizational activities. Two of the most interesting features of the work are details of very early regeneratively-cooled rocket engines—undoubtedly quite unknown to contemporary Western pioneers in this technique—and Glushko's chart of (GDL-OKB) Soviet rocket engines from 1939—1946 (pages 492—493). The chart shows that the most productive year of engine development was 1933. No engines at all show up in the 1939—1940 columns and work began again from 1941 at a very slow pace.
129. Pendray, 1955:586—587; Pendray, 1963a:124; Interview, Lasser, 1980; Interview, Lemkin, 1977; Kyle, 1976:46; "Laurence," 1961; Letter, Fierst, 1977; Letter, Van Devander, 1976.

The building at 450 West 22nd Street, visited by the author in 1976, was then being refurbished. Originally built in 1835, it was owned (though apparently not lived in) by Clement Clarke Moore, the man who wrote "T'was the Night Before Christmas." He lived across the street. A plaque had originally adorned the building but has been torn off. The Pendray apartment itself was reached by narrow stairs to the top floor. Part of the ceiling sloped and a long, narrow window overlooked the street below. The Pendrays had futuristic, cubistic furniture, to go along with their futuristic ideas. The basement apartment was then (in 1930) a speakeasy, to which several of Pendray's guests—including Robert Esnault-Pelterie, were invited. The Frenchman, Pendray recalls, was disappointed and cared only for wine. Mason's home at 302 W. 22nd St. was actually AIS headquarters.

Besides Lemkin, Manning was also foreign-born. He came from New Brunswick, Canada, and had served as a second lieutenant in the Canadian Air Force in World War I. Like Schachner, he was also a lawyer by profession but never practiced. William Lemkin may have been the earliest of the Society's founders to have met Gernsback. A chemist by profession and cartoonist by avocation, Lemkin created his character "Scienty Simon" for Gernsback's Science and Invention in 1925. Lemkin produced numerous other cartoons for publication, several of them appearing in such radio magazines as Radio News, but says he does not recall any that were space or rocket-oriented.
130. Letter, Moskowitz, 1974; Moskowitz, 1954:10; Letter, Pendray, 1976a; Letter, Pendray, 1976b; Interview, Pendray, 1977a; Interview, Weisinger, 1977; "American Interplanetary Society, Membership," ca. 1930; Kyle, 1976:135; Moskowitz, 1959:9—10.

The Scienceers were formed 11 December 1929, and met in Fitzgerald's Harlem home (Fitzgerald's address as given on the AIS membership list is 211 W. 122 St.—in Harlem). Hugo Gernsback offered to rent a room for their regular meetings at the American Museum of Natural History. According to Moscowitz, Gernsback could not attend and sent his editor, David Lasser, who went with Pendray and Lemkin. Lasser, Pendray and Lemkin allegedly pressed the Scienceers to join their Society but only one member did, Warren Fitzgerald. Michael Ashley, in his The History of the Science Fiction Magazine,

Chicago: (Henry Regery Co.: 1974). *1*:41, recounts a similar story. However, in interviews with the present author, neither Lasser, nor Pendray, nor Lemkin have any recollection of this incident, nor of The Scienceers, nor of Fitzgerald, though Fitzgerald dropped out of the AIS within a year.

That the Scienceers were space-minded is clearly shown by an editorial appearing in their magazine, *The Planet,* for September 1930: "In this age of scientific achievement, when the wonders of yesterday become the commonplace of today, the idea of interplanetary travel should be accepted as a definite possibility that is certain to be realized in the not distant future. The attitude of the press toward such projects is invariably flippant. Whenever the subject of interplanetary communication appears in the news, it is treated with unconcealed levity and ridicule. It seems to us that this attitude is wholly unwarranted. . . ."

131. "American Interplanetary Society, Membership," ca. 1930; Pendray, 1955; 586—587; Pendray, 1977b; "Lichenstein," (Misc.).

Pendray recalls that Giles wrote some fiction. He may have been related to the later ARS member and president, Cedric Giles. Lasser and Pendray cannot recall with certainty whether Roy Giles was a founder. Giles' brief obituary may be found in *The New York Times,* 1 January 1942:25:2.

132. "American Interplanetary Society Associate Members," (Misc.); Interview, Pendray, 1977b; "Lichenstein," (Misc.).

In the VfR at least one of the founding members later wrote science fiction. This was Willy Ley. The famous German science-fiction author, Otto Willi Gail, later became a VfR member and the earlier German science-fiction author, Kurd Lasswitz, exerted considerable influence upon the Society through his works.

133. Pendray, 1955; 586—587; Letter, Van Devander, 1976.

134. Pendray, 1955:586—587; American Interplanetary Society, *Constitution,* 1930.

135. American Interplanetary Society, *Constitution,* 1930; Manning (Misc.), 1931; Interview, Pendray, 1977a; Pratt (Misc.), 1931; Letter, Brandt, 1930; American Interplanetary Society, *Membership,* (Misc.) ca. 1930; Moscowitz, 1959:7; Letter, Pendray, 1976a; Letter, Pendray, 1976b.

In June 1930 a Mr. C. A. Brandt, writing on stationary from the New York Liederkranz, a choral society, offered to sell his "very complete library" to the Society, "containing everything that has ever been printed about interplanetary travels, scientific fiction, etc." Brandt's library was not bought, mainly because the AIS could not afford it and Brandt wanted only cash. C. A. Brandt was a German-born chemist who was then, according to science fiction historian Sam Moskowitz, "the greatest living authority in the world on science fiction." This statement was no exaggeration. Brandt, who among other accomplishments was responsible for the introduction of calculating machines in America, possessed a fabulous library of science fiction and fantasy, including works in the German, French, and Scandanavian languages." Moskowitz also says that "The man Gernsback leaned on most heavily for the selection of stories was C. A. Brandt."

136. Pendray, 1955:587; "News of the Society," 1930:1; Interview, Pendray, 1977a.

137. "News of the Society," 1930:1; Pratt, 1930:2; "Current," 1930:2—3; "News From Abroad," 1933a:3.

138. Goddard, Esther and Pendray, 1970:735.

139. Pendray, 1955:588; Goddard, Esther and Pendray, 1970:735.

140. Oberth, 1974:140.

141. "News From Abroad," 1930b:3; "Oberth Rocket," 1930:8; Brügel, 1933:135—137; "German Society," 1930:4.

142. "News of the Society," 1930:1; Pendray, 1955:588; "Pelterie Here," 1931:1; Goddard, Esther and Pendray, 1970:780.

143. Pendray, 1955:386—588; Interview, Pendray, 1977a; American Museum, 1931 (Misc.).

The existing bill of the American Museum shows that 32 additional attendants were hired for the show at $2.00 each, plus one motion picture operator. The total expenditures, which was then considerable, was $72.50 charged to the Society.

144. "Two Thousand," 1931:1; "To Hear," 1931.

145. Advertisement, 1932 (Misc.).

146. Mason (Misc.) 1931; American Interplanetary Society, (Misc.), ca. 1931; Pendray, 1955:587—588; "Sheldon," 1964:19; "To Publish," 1931:8; Lemkin, 1931:2—5; "The Future,: 1931:6; "Society Holds," 1931:2; Interview, Smith, 1977.

Clyde Fisher's name was prominent in early Society affairs. He too may be said to have been a Gernsback alumni as he was one of the astronomy associate science editors for Gernsback's *Science Wonder Stories* at the same time Lasser had been literary editor.

147. Pendray, 1953a:126; Ley, 1961b:140; "Has Two Step," 1931:1; "Altitude Rocket Explodes," 1931:8; "Dr. Lyon Selects," 1931:3; "Lyon Preparing," 1931:8; Letter, Pendray, 1976b; Interview, Pendray, 1977a; Interview, Pendray, 1976b; Letter, Ley, 1931c; Blosset, 1947:9.

In Germany, Willy Ley already knew of Lyon's deceit and informed Pendray later that year. On 2 November 1931 Ley told him: "You ask for Dr. Darwin O. Lyon, I'm afraid to tell you, that I have heard very much about him, but no good things. I don't like to speak before knowing more. I will only tell you (only for you personally, don't tell it to the AIS) that his former secretary has written to me and to von Pirquet (she knows him personally from Vienna) that all the stories around D.O.L. are not true. She says he never had a rocket, and never shotten a.s.o. [and so on]. It is true the successful experiments of L. he tells, were only seen by himself alone, and by no other man. I hope, I can clear it up."

148. Pendray, 1963a:126.
149. "Society Incorporates," 1931:1; Pendray, 1931a:5—12; "Letter From Dr. Goddard," 1931:9—10; Lehmann, 1963:169—170.
150. "Preliminary," 1931:4—7; Pendray, 1932:4; Pendray, "Experiments," 1931:8; "Pierce," (Misc.); Pendray, 1955:589; Pendray, 1935:9—12; Interview, Pendray, 1977b; Letter, Pendray, 1932; Letter, Pierce, 1931.
151. Pendray, 1974:142; Pendray, 1932a:3, 7; "Society's Rocket To Be Exhibited," 1932:8.
152. Pendray, 1955:589; "Society's Rocket To Meet," 1932:7—8; Pendray, 1974:2—4; Interview, Lemkin, 1977.
153. Letter, Pendray, 1978; Interview, Pendray, 1977b; "Conquest," 1931:10; "Society Elects," 1932:1; "Lasser," (Misc.); "Report of the President: 1932:7; "Rynin," (Misc.); Norton and Manning, 1932:4—6; Generales, Jr., 1974:75—76; Rynin, 1970:3(8)337—340; Interview, Lasser, 1980.
154. "Schachner," 1932:2—3; Interview, Smith, 1977; Pendray, 2955:589; "Flight of Experimental," 1933:1.

Constantine Paul Lent, later the Vice-President of the Society (1943) and noted for his own original designs, made a constructional drawing of ARS rocket No. 2 in 1935. The picture appears in the March 1936 issue of *Astronautics* and is claimed by Lent to have been the first professional engineering drawing made by the ARS.

155. "Flight of Experimental," 1933:1—11; Pendray, 1974:142.

While ARS rocket No. 4 was the last of the Society's liquid-fuel flight rockets, H. F. Pierce independently launched a liquid rocket to about 76 meters (250 feet) on 9 May 1937 at Old Ferris Point, in the Bronx. This rocket is depicted on the cover of the April 1938 issue of *Astronautics* (No. 40), though is unidentified as such.

156. "Laurence E. Manning Becomes President," 1933:12; "Treasurer's."
157. Pendray, 1974:590; "Report of Rocket," 1934:5; Pendray, 1963b:390.
158. American Rocket Society (Misc.), ca. 1934; Letter, Klemin, 1934; Letter, Pendray, 1938; Wyld, "A Program," (Misc.), 1938; Letter, Pendray, 1934; Pendray, 1955:590; Forthcoming, 1934:7; American Interplanetary Society, *Constitution* 1930; Pendray, 1974:154; Letter, Pendray, 1934a.
159. Pendray, 1955:590—591; Shesta, 1935:4; "Proving Stand," 1935:6; "Report of Motor," 1935:3—4, 20; Africano, 1936:3—5, 20; "Rocket Motor," 1936:5, 13, 20; "Motor," 1935:17; "Woman,: 1935:3; Interview, Smith, 1977.
160. "Rocket Motor," 1936:5, 13, 20; Van Dresser, 1936:8, 18; Wyld, 1934:8—13; Pendray, 1955:591—592; American Rocket Society, *Supplemental,* 1936:*passim;* Wyld, 1936:13—16; Sänger, 1936:2—121; Klemin, 1937:7—9; "Treasurer's Report," 1933:15; "Rocket Experiments," 1934:2; "Flight of Rocket," 1934:1—2, 12; "Message" 1934:1; Interview, Shesta, 1976.
161. "Notes and News," 1937:2; "Notes and News, 1938:2; Rocket Society, 1937:12; Wyld, 1938:11—12; Shesta, Pierce and Wyld, 1939:1—5.

The priority of the regeneratively-cooled rocket motor is hard to pin down. The finest study made upon this development in Europe is Dr. Irene Sänger-Bredt and Rolf Engel's 1968 International Astronautical Federation paper, "The Development of Regeneratively Cooled Liquid Rocket Engines in Austria and Germany, 1926—42," published by Frederick C. Durant, III and George S. James, editors, *Smithsonian Annals of Flight No. 10:*217—246. The partly cooled regeneratively-cooled rocket motor of the American Harry Bull is dealt with in the paper "Harry Bull, American Rocket Pioneer," by Frank H. Winter, presented at the 1976 International Astronautical Federation Congress, Anaheim, California.

162. "New Experimental," 1939:3; Williams and Epstein, 1962:196; "Rocket Demonstration," 1939:16; "Report on Motor," 1941:3—5; "Wyld Motor," 1941:8; Pendray, 1955:592; Pendray, 1974:154; Shesta (Misc.) 1978:1—16.
163. Great Britain, 1875:(17)1.
164. Interview, Clark, 1977; "England," 1931:1195; Kyle, 1976:73, 100, 120; Letter, Cleator, 1931; Letter, Cleator, 1978; Interview, Cleator, 1978; "Interplanetary Progress," 1931:1332; Goddard, Esther and Pendray, 1970:819; Letter, Cleator, 1978b.

The initials after Cleator's name, A.M.I.R.E. and others he used, A.M.I.E.T., and F.R.S.A., stand for organizations of which he was a member, respectively the Institute of Radio Engineers, the Institute of Engineering Technology, and the Royal Society of Arts.

Cleator's maverick individualism also helps explain why he was a likely candidate for championing and later leading so unappreciated a movement as space flight in the 1930s. "I was," he informs the author, "also a voracious reader of such outcasts of conventional thought as J. D. Frazer, T. H. Huxley, John M. Robertson, Andrew H. White, Winwood Reade and Voltaire, to name but a few. But the two writers whose forthright opinions and superb prose style influenced me most were my fellow countrymen Bertrand Russell and the American H. L. Mencken. With Bertrand Russell, to my great loss, I at no time had any personal contact, but between Henry Mencken and myself (in common with scores of other young literary aspirants to whom he dispensed encouragement and advice), there was maintained a lively and uninterrupted correspondence for all of 20 years, up to the time of the lingering malaise which finally ended in his death in 1956." Some of the later Cleator-Mencken correspondence from late October 1939, is found in Carl Bode, editor, *The New Mencken Letters,* (New York: The Dial Press, 1977). The early correspondence, some of which naturally deals with space flight and BIS affairs, will be published in the near future.

A picture of the radium-powered Moon Rocket in the movie "All Aboard for the Moon" may be found in *Science and Invention* 8:(April 1921): 1292. Bray also produced a similàr short semi-animated film about the same time, called "Hello, Mars!"

In addition to some of the far-flung founder members of the BIS, a supplementary sheet attached to the minutes for the BIS meeting of 3 August 1934, shows that by that date there were also "local groups [i.e., BIS branches or representatives] at Manchester, Birmingham, Eccles, [and] Farnborough."

165. Interview, Clarke, 1977; Letter, Cleator, 1934c; Letter, Cleator, 1933; "Interplanetary Society Progresses," 1930: 754; "Letter from the German," 1931: 1044; "Interplanetary Progress," 1931: 1332; "American Interplanetary Society," 1931: 134; Cleator, 1933; Cleator, 138b: 15—18.

A marvelously satirical time table of the possible history of interplanetary travel written by Cleator and appearing in the Manchester Interplanetary Society's journal *The Astronaut,* for August 1938, shows that even by that late date there were still many in England and probably elsewhere, who regarded space travel as both impossible and insane. Several excerpts from this article are worth quoting: "57 B.I.T. [Before Interplanetary Travel]. *3 May:* The International Rocket Society announces that the construction has begun of a rocket vessel capable of reaching the moon . . . *12th February:* A group of eminent professors demonstrate, to the entire satisfaction of themselves and the world, that an extra-terrestrial voyage is physically, chemically, and biologically impossible . . . *27th March:* A frantic, semi-Christian mob, ten thousand strong, wreck the completed space-ship. *5th April:* The building of the second ship is begun. 44 B.I.T. *25th March:* The second ship, shot secretly moonwards, backfires at a height of 53 miles, and crashes in flames on the British House of Lords, causing the untimely demise of five bishops eight peers of the realm, and a charlady . . . 1 B.I.T. *31st December:* Man reaches the moon! . . . A.I.T. 1. *5th January:* A group of eminent professors demonstrate . . . *27th February:* The Martian expedition effects a landing, and radios the discovery of intelligent beings . . . "

During one early meeting of the BIS, the members became so carried away with their talk of trips to other worlds that they scarcely noticed "that a most infernal din was going on in the near vicinity." The noise turned out to be the commotion caused by a fire below. Finally the meeting had to cease when it was rudely interrupted by a member of the Liverpool Fire Brigade who crashed through the door. He was soaked, said Cleator, had his "axe in hand, and unable to credit the evidence of his smoke-filled eyes."

166. Cleator, 1934a: 2—3; Allward, Carter, Ross and Gatland 1967: 150; "Membership," 1934: 5; Letter, Cleator, 1978; Interview, Burgess, 1977; Cleator, 1936a: 140, 141, 149; Letter, Cleator, 1978; Cleator, 1950: 49—51; "Editorial" 1938a: 3.

Werner Brügel's IRKA (also given as International Rocket Society or Bureau of Information) would have been headquartered in Frankfurt-am-Main if it materialized.

Brügel also contacted fellow German Ernst Loebell, founder of the Cleveland Rocket Society. Loebell subsequently published a brief notice of the IRKA in the CRS journal *Space.* Loebell added: "The Cleveland Rocket Society expects to cooperate with this organization fully." This was in 1934.

Count and Countess von Zeppelin joined the BIS in 1936 when they lived at Blackwater, Hants, England. Cleator kept up correspondence with them until the outbreak of the war. The letters to Cleator were destroyed as a result of a bombing raid in 1941. One day during the war while he was "idly tuning in to a German propaganda talk," Cleator says he "was startled to hear a member of the British Interplanetary Society [Countess von Zeppelin] discoursing on the stupidity of conflict between the British and German peoples."

167. Cleator, 1934c: 13; Letter, Cleator, 1978; British Scientists, 1934; Brügel, 1936: 6; Gartmann, 1956: 95; Letter, Ley, 1935; "Analecta," 1935b: 13; Cleator, 1950: 50, 52; "International Rocket," 1034: 34; "Mail," 1934.

168. Cleator, 1934e: 20; "Mail," 1934; "British Rocket Car," 1934: 347; Cleator, 1934d: 14; Interview, Pendray, 1977b; Cleator, 1934b: 15; Smith, "Technicalities," 1937: 8; Letter, Cleator, 1934a; Letter, Cleator, 1934b; Letter, Cleator, 1978; Pendray, 1935: 9—12.

In the Spring of 1934 Cleator wished to show portions of the German movie *Frau im Mond* to current and prospective members of the BIS, just as the American Interplanetary Society **had** done several years earlier. With some difficulty, Cleator procured the use of a theatre for a Sunday showing and also alerted the press, but at the last moment the London representatives of the Ufa film company courteously informed him that no copies of the film were to be had in England.

169. Strong, 1934: 33; Cleator, 1950: 51.

170. Cleator, 1936a: 2; "International Interplanetary," 1934: 18; "Editorial," 1935: 15—16; Stranger, 1936: 17—18, 21.

171. "Notes and News," 1936: 23; "Centerbladet," 1937: 13; Letter, Cleator, 1978; Interview, Cleator, 1978.

172. Cleator, 1935: 2; Smith, 1937: 9—10; Interview, Burgess, 1977.

173. Cleator, 1961: 114; Letter, Cleator, 1936a; Letter, Cleator, 1941; Cleator, 1948: 97—99.

174. "Herr Zucker's," 1934: 27; "British Rocket Mail," 1934: 8; Ellington and Zwisler, 1967: 84—85; Letter, Pendray, 1934b; "Gerhard Zucker," (Misc.); Harper, 1946: 37; "The Four," 1934; *"Flug,"* 1936: 80.

175. "Death Ray," 1973; "Briton," 1938: 24; "Wire-Tailed," 1943: 8—9; Letter, Cleator, 1936a; Ross, 1950: 98.

Grindell-Matthews was often called "Death-Ray" Matthews because, from the early 1920s, he promoted a high-intensity electromagnetic wave transmitter whose beams could allegedly explode gun powder at a distance, kill small animals, cure certain diseases, stop automobile and aircraft engines, and blow up stores of explosives and ammunition, etc. The Matthews "death ray" was envisioned by its inventor and the sensationalistic press as a potential terror weapon. Russian astronautical writer, N. A. Rynin, included it in the "Radiant Energy" volume of this *Interplanetary Flight* encyclopedia. See N. A. Rynin, *Interplanetary Flight and Communication* (NASA TT F-642), 1(3):17—24.

176. Ley, 1961b:178—180.

177. Allward, Carter, Ross and Gatland, 1967b:201; "Correspondence," 1936:32; Letter, Cleator, 1936d; Letter, Cleator, 1936c.

178. "London," 1937:18—19; "Profiles," 1969:41.

179. "London," 1937:13, 19; "Editorial," 1937a:3—4; "Editorial," 1937b:3; Ross, 1950:95—96.

180. "Another," 1936:22, 34; Interview, Clarke, 1977; Goddard, Esther and Pendray, 1970:1026; Gifford, 1971:140; Martin, 1927:29; Bloom, 1958:178; Cleator, 1950:52; "Analecta," 1935a:5; Mayer, 1937:4—5.

Low's enthusiasm for interplanetary flight extended back considerably before the BIS was founded. During World War I he received a patent for a radio-controlled air-to-air rocket and also worked on radio-controlled pilotless planes which are considered amongst the world's first guided missiles. Low's story is related in Ursula Bloom's *He Lit the Lamp—A Biography of A. M. Low* (London: Burke Publishing Co., 1958).

H. G. Wells was quite cognizant of the BIS and of the astronautical movement in general. He had had some correspondence with the American Goddard but was reluctant to join any society.

181. Advertisement, 1935:5; Interview, Burgess, 1977; "Boy," 1937.

182. Interview, Burgess, 1977; Burgess, 1944:6—10; "By Rocket," 1936; "Young,: 1937.

183. Interview, Burgess, 1977; Cleator, 1938a:3—6.

184. Interview, Burgess, 1977; "Biography," 1941:3—6; Allward, Carter, Ross and Gatland, 1967c:234; Burgess, 1944:10; Interview, Traux, 1978.

Truax and his fellow shipmates aboard the *USS Wyoming* were not permitted to take shore leave in Spain which was then in its Civil War; they merely stopped at the port of Santander. In any case, Truax says he had no time at all to investigate astronautical or rocketry activities at any of the other places during his cruise.

185. Smith, 1942:d—f; Ross, 1969:42—47; Dooling, 1974: 82—84; Ross, 1974:209—216.

186. Kyle, 1976:137; Ross, 1950:97—99; Interview, Truax, 1978; Ross, 1984:209—216.

In 1936 Maurice K. Hansen started the British amateur science fiction magazine *Novae Terrae* in behalf of Hugo Gernsback's Science Fiction League which had branches in England. It was the official organ of the British Science Fiction Association and was edited by Ted Carnell, also a BIS officer.

197. Ross, 1974:209—216.

188. Ross, 1974:209—216.

189. Ross, 1974:209—216; Ross, 1950:100—101.

190. Letter, Cleator, 1939; Letter, Cleator, 1963; "Letter," 1939:10; "Profiles," 1969:40.

191. "Advances," 1928; "Flight to Venus," 1928; Uhler, 1978:78—79; "All Aboard," 1928:19; "From the World," 1928:356; "Plans," 1928; Goudket, 1935:14—15.

Perhaps Condit should not be judged too harshly. Harry B. Uhler, a fellow experimenter, recently published his recollections of Condit and his Venus rocket in *Science News,* 114 (29 July 1978):78—79. He recalled Condit as a "mathematical genius" who directed the construction and testing of a 7.3 meter (24-foot) long rocket powered with 189.3 liters (50 gallons) of gasoline which was vaporized and ignited by a spark plug. The ship was allegedly loaded with the fuel with Condit in it. Upon the first and only known attempt to take off in the craft, in 1928, huge flames and smoke gushed out of the rear but the ship failed to move. Condit's science was surely lacking, but not his bravado and his then innovative use of gasoline as a fuel several years before Goddard's experiments were known to the public.

192. Letter, Prindle, Jr., 1934; Andres, 1934:30—31; Letter, Poggensee, 1977; Pendray, 1945:115; Letter Loebell, 1978a; Letter, 1979; Interview, Loebel, 1978; Letter, Schofield, 1965.

The history of the Cleveland Rocket Society presented here is by no means a complete one. In 1964 John Tascher presented a 75 page dissertation on the Society for his Master of Arts degree at the Case Institute of Technology in Cleveland. In addition, there is a great deal of original material in the Cleveland Rocket Society Collection within the Archive of Contemporary Science and Technology at Case.

193. Interview, Loebell, 1978; *Cleveland,* 1930:177; Letter, Loebell, 1978a; Tascher, 1955:49—50; Hanna, 1933b:2, 8; Letter, Hanna, 1964a; "Ohioans," 1933.

194. Letter, Prindle, Jr., 1934; Tascher, 1966:50—51; Letter, Cleveland Chamber, 1933; Interview, Loebell, 1978; Letter, Hanna, 1964a; Loebell, 1935:13, 18; Prindle, 1934:3.

195. Tascher, 1955:50—51; Letter, Hanna 1964a; Letter, Prindle, Jr., 1934; Spangenberg, 1934:4—9, 12.

196. "Lt. Comdr.," 1936:18; Interview, Settle, 1977; Letter, Hanna, 1964a.

197. Interview, Reeb, 1978; "Rocket Model," 1934:3; "Cleveland Gets," 1935; "Complimentary," 1934—12; Letter, Hanna, 1964a; "Cleveland Rocket Society Extends," 1933c; Hanna, 1933a:a; "Auf Wiedersehen," 1933:b.

CRS publicity included a display of a rocket in the window of the Reeb Drug Store, and another display in the Higbee Department Store. It was advertised, but never realized, that a CRS rocket might set a new record in the Cleveland Air Races that annually drew 50—100,000 people.

198. "Manned," 1933:2; Ley, 1937:534—535; "Trophy," 1933:3; "Attend," 1933:2; "Interesting, 1934; "By Rocket to the Moon Shown," 1933:1; Letter, Hanna, 1964a; "Rumor," 1933:3; Philip, 1937:88—90; McNash, 1935:150, 182.

The apparent "approval" and propagation of the Fischer and Darwin Lyon hoaxes were, unfortunately, amongst the negative roles of some of the societies. The hoax-side of the astronautical movement also underscores the low state-of-the art and naïveté during this period.

199. "Constitution," 1933:4—7; "CRS Constitution," 1934:8; Letter, Hanna, 1964b; Letter, Hanna, 1964a; Tascher, 1966:51.

The name "Proving Field Laboratory" seems a Germanic influence, the German words for "test" or "testing" being *"Prüfung"* and *"Probe."*

200. Waterhouse, 1934:300, 336; "Plans Rocket," 1934:41; Interview, Settle, 1977; "Society Offered," 1934:6; Tascher, 1966:53; Letter, Loebell, 1979.

201. Tascher, 1966:53—55; "Two Firing," 1933:9; Letter, Hanna, 1964b; Letter, Prindle, 1934; Waterhouse, 1934:336; Letter, Loebell, 1979; "New Cachet," 1935:20; Advertisement, Cleveland, 1934:21; Nebel, 1972.

Probably the first society that promoted mail rockets was the VfR, in 1931, although Franz von Hoefft of the Austrian Society for High Altitude Exploration presented his ideas on intercontinental mail rockets in the VfR journal *Die Rakete* in 1928. In 1931 the VfR's Rudolf Nebel spoke of the possibility of a postal rocket flying from Berlin to Munich in 10 minutes. He took his plans further and contacted the State Post Ministry. Shortly after, four of the Ministry's directors witnessed a demonstration of a rocket at the *Raketenflugplatz.* The leader of this delegation, a Professor Kuckkuck, exclaimed that he still had "no understanding of the reasons for the recoil of the rocket in the back or front," but was impressed with the promise of rocket power nonetheless. Consequently the Post Ministry sent an honorarium to the VfR in support of the work.

202. Tascher, 1966:55—58; Letter, Hanna, 1964a; Letter, Hanna, 1964b; "Seek," 1934; "In League," 1961:14; Letter, Loebell, 1979; Interview, Burke, 1980.

203. Tascher, 1966:60—61; "Cleveland Experiments," 1936:25.

204. Tascher, 1966:60—61; "Harvey Firestone, Sr.," 1934:21; Letter, Ley, 1952; *Cleveland,* 1935:916; Interview, Loebell, 1978; Letter, Hanna, 1964a.

205. Tascher, 1955:61—62; Interview, Loebell, 1978; Letter, Loebell, 1978a; Letter, Lencement, 1937; Letter, Lencement, 1938; "Cleveland Rocket Society," 1937:19; Exposition Internationale, (Misc.) 1937; *"L'Astronomie,"* 1937:185—188; Lencement, 1937:20—22.

Asked for details of Lindbergh's witnessing Loebell's rocket exhibit, Loebell informed the author that: "I never met nor corresponded personally with Mr. Charles Lindberg [sic], but received an account of his visit in a letter from the 'Le Commisaire General,' whose name I am unable to decipher."

206. Letter, Kaiser, 1939; Interview, Loebell, 1978; O'Donnell, 1958; Wellman, 1944; "William P. Lear," (Misc.); Andres, 1934:31; Letter, Cleveland Public Library, 1978.

207. Malina, 1973:442—456; Malina 874:113—127; Malina, 1977a:153—201; Malina 1977b:339—383; von Kármán, 1967:234—267; Letters, Malina, 1936—3946:4, 9.

Malina says he was initially inspired in space travel at the age of 12, in Czechoslovakia, the homeland of his parents where he read the Czech language edition of Jules Verne's *From the Earth to the Moon.* On his return to the United States in 1925, he continues, "I followed reports on rocket work as they appeared from time to time in popular magazines." From 1930 this reading probably included AIS/ARS developments. An example of ARS publication of GALCIT's work is found in the ARS journal *Astronautics* for July, 1938:3—6.

208. Giles, 1944:12—12.

209. Giles, 1944:12; "Our Journal," 1934:1; Letter, Kaiser, 1939; *Polk's Peoria,* 1934:177: "Studying," 1936; "History in Brief," 1939:1.

210. "Our Journal," 1934:1; "Our Policy," 1939:2; Davis, 1939:2; Look, 1939:3; "Mars," 1939:3; "Support," 1939:4; Letter; Kaiser, 1939; *Kurze Berichte,* 1939b:49; Giles, 1944:13.

211. Williamson, 1965:16—18; Interview, Williamson, 1979; Letter, Schiff, 1978; "Yale Clubs," 1937.

212. Williamson, 1965:16—18; Interview, Williamson, 1979; Interview, Gates, 1979; "University," 1938:7; "Rocket Society," 1937:12.

213. Interview, Williamson, 1979; Williamson, 1965:16—18; Interview, Gates, 1979; Gates and Williamson, 1937:3, 20—21; Sänger, 1936:2—12.

214. Williamson, 1965:16—18; Interview, Williamson, 1979; Interview, Beattie, 1979.

215. Interview, Moskowitz, 1977; "President William S. Sykora," 1936:4; "Branch Club News," 1935:1; Sykora, 1935:1; Sykora, 1936:8—11.

216. Ellington and Zwisler, 1967:212—213; "Mail Rockets," 1935.

217. Goddard, Esther and Pendray, 1970:906—907; Sykora, 1935:1; Sykora, 1936:8—11.

218. Giles, 1944:13; Limber, 1937a:4—5, 30—32; Limber, 1937b:4—5, 54—56; Limber, 1938a:12; "Letters," 1938:17; "New Experimental," 1939:3; Limber, 1938b:2; "Officers," 1938:4.

219. Giles, 1944:12; Wellman, 1944; "Seek," 1934; McNash, 1935:150, 182; McNash, 1935:26, 58; Goddard, Esther and Pendray, 1970:1007—1008.

220. Giles, 1944:13; *Clifton,* 1956:403; "Rocket Society Affiliates," 1937:12.

221. Giles, 1944:12—13; "California," 1943:13; Smith, 1940:10—11, 15; "News of the Month," 1932:4.

222. Winter (Misc.), 1976:20; Haley (Misc.).

223. Interview, Schaefer, 1977; Winter, 1977:247; Letter, Matarazzo, 1934; Letter, Matarazzo, 1935; Letter, Schaefer, 1936; *Volanzan,* 1932; Letter, Tabanera, 1979a; Letter, Tabanera, 1979b.

The sudden demise of Matarazzo's group is summed up briefly: Matarazzo abandoned his studies at the University and began to work at his father's macaroni factory.

Teófilo Tabanera, founder of the Asociación Astronáutica, Argentina, in 1949, himself became interested in interplanetary flight in the early 1930s, but had no knowledge of Matarazzo, then and in subsequent years. Two reasons are that in the 1930s Tabanera was a student at La Plata University, away from Buenos Aires. Secondly, Matarazzo's "group" had all but vanished soon after it was formed. Interest in space travel was still high in Argentina during these years, as evidenced by several space and rocketry items appearing in the influential magazine *Aeronáutica Argentina,* for the single year 1936.

The names of the Argentinians Ezio Matarazzo and Adelqui Santucci are decidedly Italian. This is explained by the fact that from the 1880s to the early 1900s Argentina received many Italian and other European immigrants because of the need of more manpower to raise cattle as a result of the introduction of refrigeration.

224. Giles, 1944:13; Ellington and Zwisler, 1967:87—95, 97, 98a; Allward, Carter, Ross and Gatland, 1967c:201; Interview, Burgess, 1977; Nebel, 1962:114; Kaiser, 1939b:25—27; "Kurze Berichte," 1939a:14; "Analecta," 1935a:5; "Miscellaneous," 1934:8.

225. Houtman (Misc.), 1930's; Giles, 1944:13; Letter, Thoolen, 1935:v.B.; 1935:124—127; "Prof." 1949:8—9; Ellington and Zwisler, 1967:1; Letter, Falkenberg, 1931; Letter, de Koningh, 1948; Field, 1937:19.

For details of some Australian Rocket Society mail flights, see "Australian Rocket Experiments," in *The Indian Air Mail Society Quarterly Bulletin,* 11 June 1937:50—51. The astronautical or rocketry movement reached Australia earlier. In 1931, for example, one B. Falkenberg of "Bonnie Hills," Byaduk, Virtoria, Australia, wrote to the Secretary of the American Interplanetary Society desiring membership and briefly recounting his own crude experiments with rocketry. Falkenberg also expected "to see people flying to the moon in a few years' time."

216. Lencement, 1937:20—21; Interview, Burgess, 1977; Harmon, 1931:205—206; Ananoff, 1931:463—466; Ananoff, 1933:409—410; Ananoff, 1935:419—437; Ananoff, 1937a:225:228; Ananoff, 1937b:271—278; "L'Astronomie," 1937:185—188; Ananoff, 1978:72—73; Letter, Ananoff, 1938; "Nouvelles," 1938:237; Ananoff, 1969—v—vi, 325—331; "Kurze Bericht," 1939a:14; Kaiser, 1939a:8—9; Ananoff, 1940:3—6.

227. Letter, Ananoff, 1938.

228. Letter, Ananoff, 1938; Rynin, 1971:2(4)71; Ananoff, 1978:72—75, 113; "Communication," 1942:15; "Communication," 1945:21; "Alexandre," 1950:41; "Notes and News," 1950:137.

For all of his work in furthering the cause of astronautics, Ananoff was awarded the first Hermann Oberth Medal in 1950.

229. "From Tokyo," 1935—2; Acme (Misc.), 1934; "A Japanese," 1935:18; Ananoff, 1937b:227; "Rocket Artillery,: 1934; Letter, Koizmu, 1931a; Letter, Koizmu, 1931b; Goddard, Esther and Pendray, 766—767; U.S. Army, 1946:4.

An item in *Astronautics* for February 1943 says: "The Japanese showed considerable interest in the work of the American Rocket Society prior to the outbreak of the war. The Imperial Army had a standing subscription to *Astronautics.*" In the U.S. Army's Technical Center's *Survey of Japanese Rocket Research and Development,* completed in January 1946, it says that "Prior to 1930 no research, theoretical or practical, government or private had been done in Japan on rockets. In consequence early investigations depended entirely upon German and English literature on the subject. The first experiments were conducted in 1931 at the First Army Technical Research Laboratory and were confined to research on propellant powders to determine a suitable composition and shape . . . In 1935 the first rocket was designed." It is also known that in September 1931 the AIS and Goddard received general inquires about rockets from the New York offices of the K. M. Okura Company of Tokyo and elsewhere. This is all we know of pre-war Japanese civilian and military rocketry.

232. Interview, Stehling, 1977; Gail, 1928:52—64; Letter, Constantinescu.

According to Ingemar Skoog, the Swedish writer of the history of rocketry in his country, there were neither astronautical nor rocketry societies in Sweden or in the rest of Scandanavia in the 1920s and 30s. Interest was there, however. Otto Willi Gail's *Mit Raketenkraft in Weltenall* (The Rocket Force Into Space) was translated into Swedish as *Med Raket Genom Varldsrymden (Tryckeri Aktiebolaget Thule: Stockholm, 1928),* with coverage of the VfR and Austrian groups. Alexandre Ananoff, in his survey of world rocketry in the 1937 *Bulletin de la Sociétié Astronomique de France,* also says that: "In Denmark there exists some constructed models and we will soon be able to establish their propulsive quality." In historical survey papers read before the International Academy of Astronautics, there is no evidence of Societies existing also in Switzerland, Spain, Italy, and Poland, in the 1920s—30s. For other activities in these countries, consult: *History of Rocketry and Astronautics: Proceedings on the Third Through the Sixth History Symposia of the International Academy of Astronautics.* Cargill Hall, editor (NASA Scientific and Technical Information Office, Washington, D.C., 1977), Vols. I and II. See also Frederick C. Durant, III, and George S. James, editors: *First Steps Toward Space* (Smithsonian Annals of Flight Number 10. Smithsonian Institution Press, Washington, D.C., 1974.) In Canada there were the makings of an earlier society or club that Stehling's. This was Clinton Contantinescu's Lethridge Scientific Society of Alberta which held "interplanetary lectures" in 1931; but so far as is known the talks were only part of their program.

231. Bainbridge. 1976:1, 45—124.

232. Bergaust, 1976:34—35; Dornberger, 1958:20, 27, 47.

233. Von Braun, 1932:449—452; Von Braun, 1930—1931:89—92.

234. Corn (Misc.), 1977.
235. Interview, Smith, 1977.
236. Bainbridge, 1976:147.
237. Letter, Deisch, 1931.
238. Letter, Zoike, 1978.
239. Ley, 1961b:445; Letter, Tabanera, 1979; Haley, 1958:269—291; Sternfeld, 1959:337; Durant, III, 1961:23, 25, 26, 28, 30, 31—39, 45, 50, 52—53, 55—58, 61, 65, 67—68, 71, 76, 78—80, 84, 88, 91.
240. Goddard, Esther and Pendray, 1970:485.

1. Pre-20th century spaceflight romance and philosophy. Cyrano de Bergerac rising to the Moon via dew drops, 1656.

2. *From the Earth to the Moon,* by Jules Verne (1865), was a later classic. It was more scientific and also inspired the great pioneers Tsiolkovsky, Oberth and Goddard.

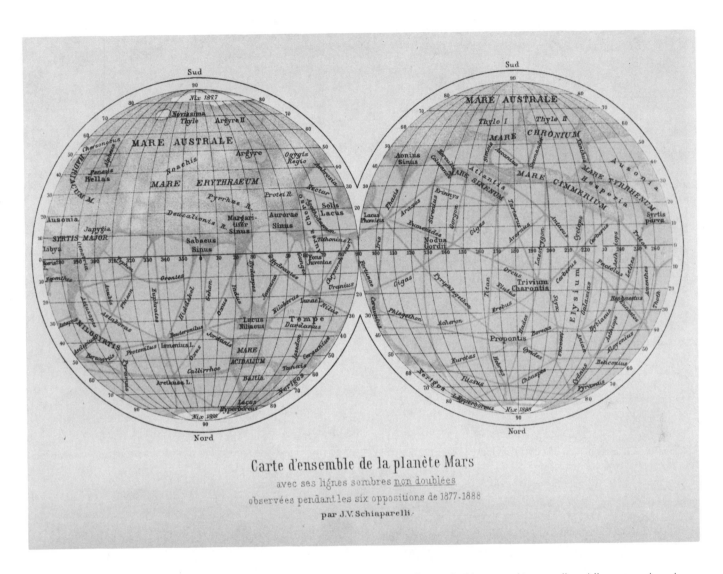

Carte d'ensemble de la planète Mars
avec ses lignes sombres <u>non doublées</u>
observées pendant les six oppositions de 1877-1888
par J.V. Schiaparelli.

3. Giovanni Schiaparelli's observations of Mars in 1877. He saw ''*canali*'' or ''channels,'' a word misinterpreted to mean ''canals''—man-made, or by intelligent beings, i.e., Martians.

"HUGE BLACK SHAPES GROTESQUE AND STRANGE"

4. Speculation abounded on what Martians looked like: a scene from H. G. Wells' *War of the Worlds* (1898). More than ever man had to explore the planets—especially Mars.

5. By the 1920's technology caught up with science fiction—almost. Science fiction directly spurred many to form societies to seriously study interplanetary flight.

6. Science fiction movies played a role too. Hermann Oberth was technical director of the 1929 silent movie *Frau im Mond* (Woman in the Moon). The movie was a favorite of the now flourishing space travel movement.

Up, Up, Up to the Distant Stars by Rocket—A Conception of the Start of a Journey by Rocket-Airship, a Project of Herr Max Valier, the Distinguished German Aeronaut, Who Hopes to Hurl a Hermetically Sealed Plane Through the Ether at 420 Miles a Minute.

7. Newspaper Sunday supplements also lavished attention upon the growing movement. The 13 January 1929 *American Weekly* featured Max Valier's hermetically-sealed "ether" plane at 675 km/min (420 miles a minute).

8. The Pioneers: Soviet "Father of Cosmonautics" Konstantin E. Tsiolkovsky, with aviation-minded grandchildren, in his backyard, Kaluga, 1934.

9. Hermann Oberth, author of the milestone *Die Rakete zu den Planetenräumen* (The Rocket into Planetary Space), 1923: the first comprehensive work on all phases of spaceflight.

10. Max Valier, publicist and experimenter. Valier wrote his own books, helped start the German Rocket Society, and was killed (1930) while experimenting with one of his motors for a rocket car.

11. Franz von Hoefft, founder of the Austrian Society for High Altitude Exploration, 1926. A leading theoretician of spaceflight who designed transcontinental and planetary mail rockets.

12. Johannes Winkler, first president and fellow founder of German Rocket Society (VfR) from 1927. Also editor of VfR's journal *Die Rakete* and builder of some of Europe's first flying liquid-fuel rockets.

13. Fridrikh Tsander, tireless Russian promoter of spaceflight and one of founders of Moscow Group for the Study of Reactive Motion (MosGIRD). Crater on the Moon named in his honor.

2-й год издания.   **Цена 20 коп.**   Пролетарии всех стран, соединяйтесь

ТЕХНИКА и ЖИЗНЬ

(КРАСНЫЙ ТРАНСПОРТНИК)

ПОПУЛЯРНЫЙ ТЕХНИЧЕСКИЙ и ПОЛИТИКО-ЭКОНОМИЧЕСКИЙ ДВУХНЕДЕЛЬНЫЙ ЖУРНАЛ

6 июля   **№ 12**   **1924 г.**

ОРГАНИЗАЦИЯ в С.С.С.Р. О-ВА МЕЖПЛАНЕТНЫХ СООБЩЕНИЙ.

Тот перелом, который наметился в последнее время в вопросе межпланетных сообщений, связанный прежде всего с работами Циолковского, Оберта и Годарда и означающий, что межпланетные сообщения из области фантазии переходят, наконец, на реальную почву,—этот перелом отразился, конечно, и в СССР. В середине апреля при Военно-Научном Обществе Академии Воздушного Флота организовалась Секция Реактивного Двигателя, которая поставила себе следующие цели:

Секция междупланетных сообщений Ак. Возд. Флота.
Каперский, Резунов, Лейтейзен.

1) Об'единение всех лиц, работающих в СССР по данному вопросу;

2) получение возможно полной информации о происходящих на Западе работах;

3) распространение правильных сведений о современном состоянии вопроса межпланетных сообщений, и, в связи с этим, издательская деятельность;

4) самостоятельная научно-исследоват. работа и в частности, изучение вопроса о военном применении ракет.

Секцией поставлен был для своих членов ряд докладов,—в том числе

Обложка первого Советского журнала, посвящен. междупланетным сообщениям.

пр. Ветчинкина и инж. Цандера; намечен конкурс на рассчет небольшой ракеты на 100 верст; создается кружок для более углубленного теоретического изучения вопроса; организуется лаборатория; открыт книжный киоск для удовлетворения наблюдающегося широкого спроса на литературу; выделена киногруппа, которая разрабатывает в настоящее время сценарий фильмы и т. п.

Секция приняла деятельное участие в организации Общества Межпланетных Сообщений. Первым шагом Общества было устройство публичного доклада М. Я. Лапирова-Скобло в Политехническом музее в Москве. Огромный успех доклада достаточно говорит о том, насколько велик интерес к вопросу межпланетных сообщений.

1-го июля предполагается выпуск первого номера журнала «Ракета»—органа ОМС.

Общество Межпланетных Сообщений временно помещается при Обсерватории б. Трындина (Москва, Б. Лубянка, 13).

1

---

14. First Russian space travel society was Air Force Academy sponsored "Section of Reaction Engines." 1924. *Tekhnika i Zhin* (Technology and Life), July 1924, ran front page story of group with founders (left to right): Kaperskiy, Rezunov, and Leiteisen.

15. *Goldnen Zepter* (Golden Scepter) ale house on Schmiedebrücke Street, Breslau, Germany (now Wroclaw, Poland) was site of founding of largest and most influential society, the VfR, 1927.

16. Leading members of the VfR, 5 August 1930, after successful firing of *Kegelduse* motor at *Chemische-Technische Reichsanstalt,* Left to right: Rudolf Nebel, Franz Ritter, unknown, Kurt Heinisch, unknown, Hermann Oberth, unknown, Klaus Riedel, Wernher von Braun, and unknown.

17. David Lasser, founder of the American Interplanetary Society, 1930. Lasser was then managing editor of *Science Wonder Stories*. Later wrote his own stories and *Conquest of Space* (1931), first book on subject in the English language.

18. By 1930 also, Germans had their *Raketenflugplatz* (Rocket Flying Place) in Berlin where the VfR planned, built and flew most of their early rockets.

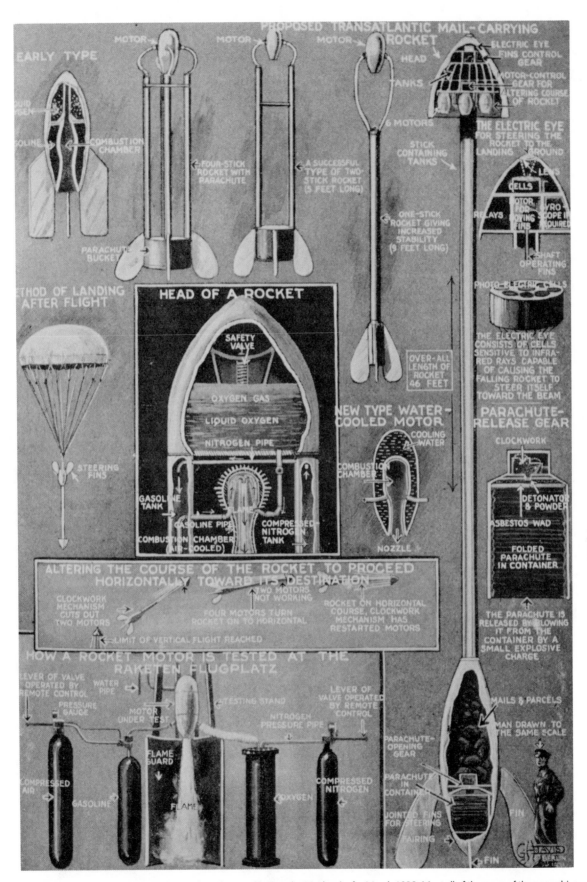

19. Evolution of German Rocket Society rockets, featured in *Popular Mechanics* for March 1932. Most all of them are of the nose-drive design and also utilize sharply swept tail fins having parabolic leading and trailing edges. The rocket societies helped introduce streamlining in rockets, though aesthetics rather than aerodynamics dictated the designs.

173

RAKETENFLUG

Aufruf!

Seit Jahrzehnten arbeitet die deutsche Wissen=schaft und Technik an dem Raketenproblem. Endlich sind wir so weit, daß greif=bare Erfolge vorhanden sind. Zur Weiterführung und zum Ausbau der Er=rungenschaften fehlt uns, die wir uns mit den klein=sten Mitteln bisher geholfen haben, das Geld. Das Ausland hat, in dem Be=streben uns unsere bis=herigen Erfolge zu entreißen, ungeheure Anstrengungen gemacht. Dies zu ver=

hindern, muß jedem Deut=schen am Herzen liegen. Möge jeder nach seinen Verhältnissen hierzu einen Beitrag geben, damit uns die Früchte jahrzehnte=langer mühevoller Arbeit nicht entgehen. Deutsch=land wird durch die Lösung des Raketenproblems min=destens in wirtschaftlicher und kultureller Beziehung derartige Vorteile erlangen, daß mit einem Schlage seine frühere Weltgeltung wie=derhergestellt wird.

Raketenflugplaß Berlin

Verein für Raumschiffahrt E. V.

Leiter: Dipl.=Ing. Rudolf Nebel, Reinickendorf=West, Tegeler Weg

Telefon: D 9, Reinickendorf 4617

Postscheckkonto: Berlin 61591
Raketenflugplaß Berlin

20. Money is always a problem, Nebel characteristically producing a jingoistic handbill calling for donations for ''our rocket work'' which ''has been taken from us by other nations.'' Scene from *Frau im Mond* adds to the appeal.

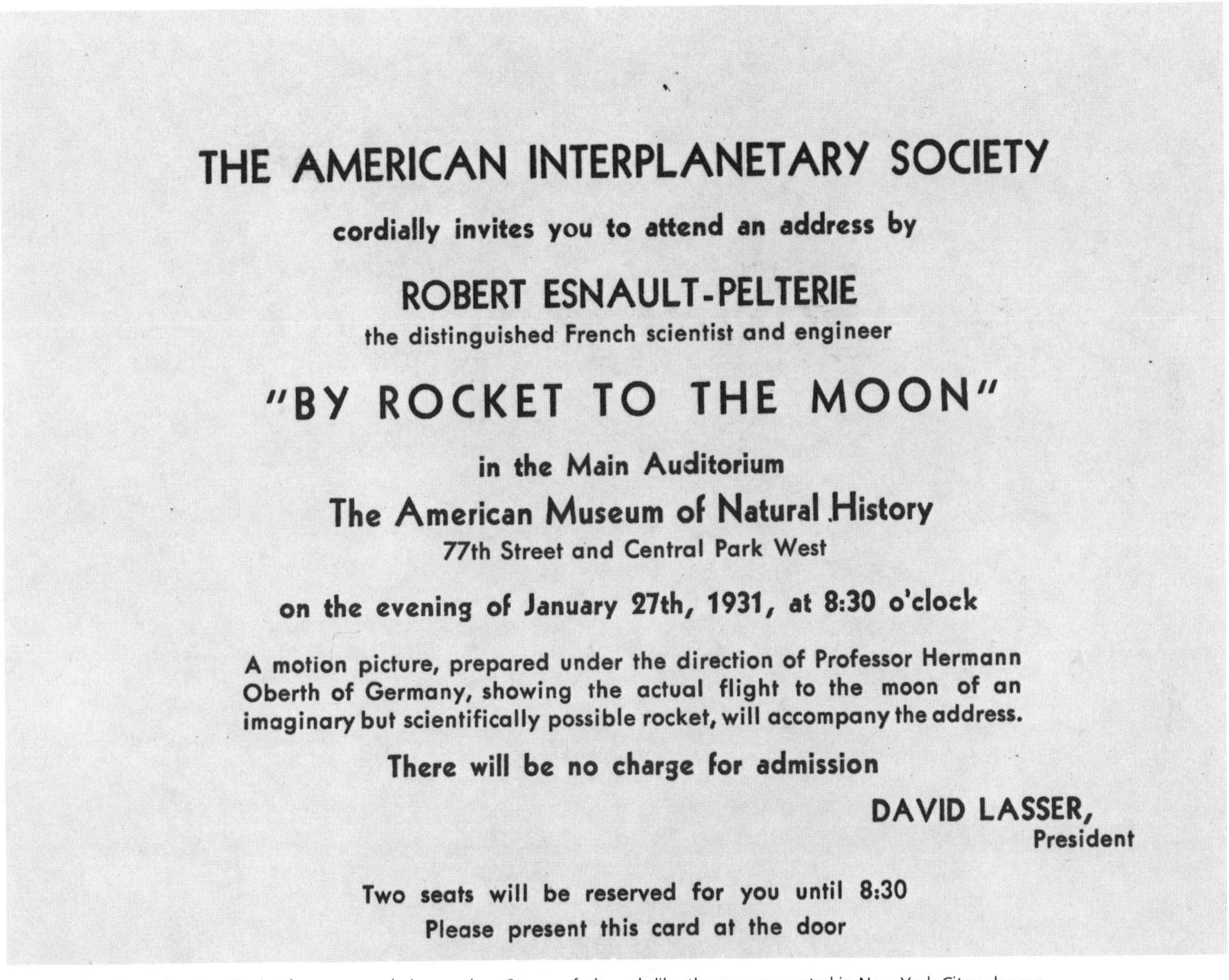

THE AMERICAN INTERPLANETARY SOCIETY

cordially invites you to attend an address by

ROBERT ESNAULT-PELTERIE

the distinguished French scientist and engineer

"BY ROCKET TO THE MOON"

in the Main Auditorium

The American Museum of Natural History

77th Street and Central Park West

on the evening of January 27th, 1931, at 8:30 o'clock

A motion picture, prepared under the direction of Professor Hermann Oberth of Germany, showing the actual flight to the moon of an imaginary but scientifically possible rocket, will accompany the address.

There will be no charge for admission

DAVID LASSER,
President

Two seats will be reserved for you until 8:30
Please present this card at the door

21. American Interplanetary Society's propaganda innovation. Scores of placards like these were posted in New York City subways.

22. Phillip E. Cleator, Liverpool construction contractor, wakens England to the trend and forms British Interplanetary Society in 1933. Picture taken that summer, at opening of Liverpool's Speke Aerodrome.

23. Orbiting spacecraft graces cover of VfR's journal *Die Rakete*.

ASTRONAUTICS

Journal of the American Rocket Society

Number 37   July, 1937

Courtesy of Science Service

Number One Man of Rockets—A biographical sketch of Dr. Robert H. Goddard. **Rocket Motor Efficiency**—What can we expect of the motor?

24. Robert H. Goddard is honored in the American Rocket Society's journal *Astronautics.*

No. 1. JANUARY, 1934. Vol. I.

Journal of the
British Interplanetary Society.

———

34, OARSIDE DRIVE,
WALLASEY, CHESHIRE, ENGLAND.

25. Futuristic spaceship designed by Cleator was winning design in contest for first cover of *Journal of the British Interplanetary Society.*

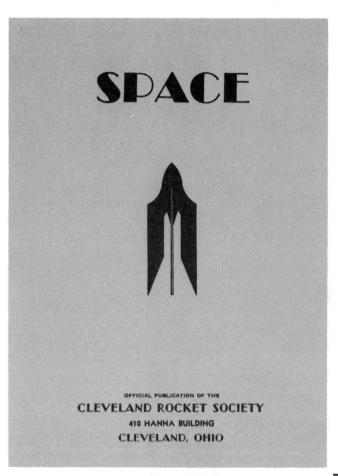

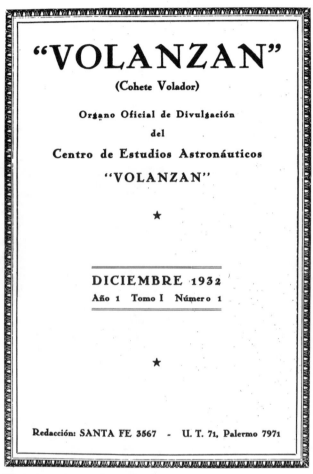

26. Lasting for five issues, from December 1933 to September 1934, *Space* was the official organ for the Cleveland Rocket Society.

27. In Argentina too there existed an astronautical journal. It was very short-lived, and this may have been the only issue.

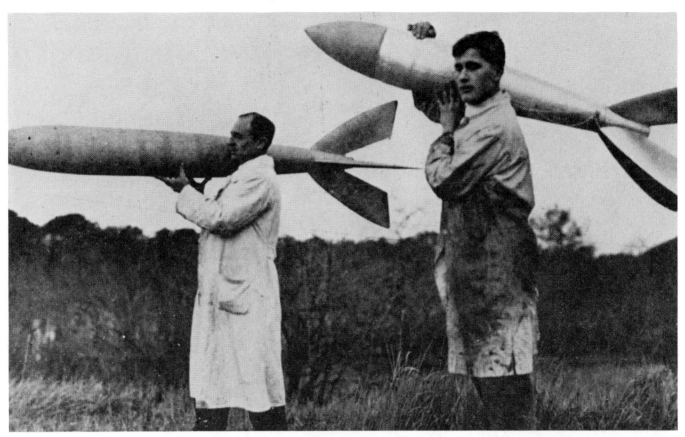

28. Nineteen or 20-year old Wernher von Braun, right, bearing a streamlined *Repulsor* rocket, follows a determined Rudolf Nebel at *Raketenflugplatz*, about 1932.

29. Pouring in the liquid-oxygen in an asbestos suit. The VfR rocket is a 10-liter (2.6 gallon) model.

182

30. Cutting corners, the Americans tip their ''lox'' from discount-purchased 15-liter (3.9 gallon) Dewar flask into coffee pot which is then transferred to rocket. Stockton, New Jersey, 12 November 1932. Coffee pot now in National Air and Space Museum.

31. Final adjustments on the Americans first rocket (ARS No. 1) are made by G. Edward Pendray inside shed at Stockton. Mrs. Pendray (Leatrice Gregory) stands by while one of the rocket's designers and builders, Hugh F. Pierce, warms hands.

32. Society's President Lasser (left), and future President, Laurence Manning, another science fiction writer, poise gingerly behind home-made sandbags for pre-Space Age countdown.

33. The Germans wait too. Riedel is on the right.

34. Pushing the switch for launching American Interplanetary Society Rocket No. 1 at Stockton, New Jersey, 13 November 1932, is Hugh Franklin Pierce. The rocket failed to fly.

35. But Rocket No. 2 did, at Marine Park, Great Kills, Staten Island, New York, 14 May 1933. Bernard Smith, left, dashes back to safety barricade after personally lighting rocket with gasoline torch.

36. VfR's flight of four-stick Repulsor. As a nosedriven rocket, the flame is barely visible.

37. Meanwhile, unknown to the West, the Russians plan their own spectaculars. MosGIRD members, 1931. Sergei P. Korolev, later to build Sputnik launchers, is second on left. Friedrikh Tsander standing third from right. Others are: Sumarokov, first on right; Yuri Pobedonostev next; glider designer B. I. Cheranovsky, middle with cigarette; Zaborin to his right, and A. Levits, standing.

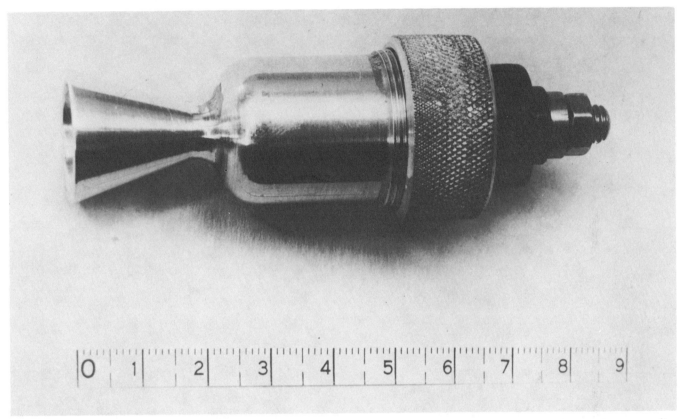

38. Electric rocket motor was initial project of Soviet Union's Gas Dynamics Lab, Department II, built and tested 1929— 1933. It electrically vaporized liquid solid ejection mass. Preliminary work on true electric propulsion, utilizing charged particles, was conducted before World War I by Robert H. Goddard and students. Department II also conducted first official work on liquid propulsion for USSR.

39. GIRD-X, Soviet Union's first liquid rocket, 25 November 1933, in forest of Nakhibino, outside Moscow. Technicians are generally not identified, possibly because of security reasons or purges of 1937—1938. Left is Korolev, team leader. Others are believed to be: L. S. Dushkin (glasses); opposite to him, at right of rocket, standing, L. I. Korneyev. Woman may be L. N. Kolbasicha. Crouched in middle row, far right, may be A. I. Polyarny.

40. Soviet GIRD 09 rocket being fueled with liquid oxygen prior to first launch, 17 August 1933. This was USSR's first hybrid rocket, the oxygen burned with solidified gasoline. Left to right: Sergei Korolev, GIRD head; Nokolai Efremov, senior engineer, GIRD second team; and Yuri Pobedonstsev, GIRD third team head.

41. GIRD 09 flies to 400 meters (1,312 ft.). Other 09 models are built, reaching up to 1,300 meters (4,165 ft.). These flights were not made known to the West at the time.

42. If the Russians failed to share their knowledge, the Germans did not. American G. Edward Pendray (center with beard) visits *Raketenflugplatz*, April 1931. Willy Ley and Rudolf Nebel to his right, at "proof stand" with valves opened by pulleys.

43. Schematic of the German Rocket Society's static rocket test stand at the *Raketenflugplatz*, Berlin, 1932, drawn by *Popular Mechanics* special correspondent G. H. Davis who visited the *Raketenflugplatz*. The stand framework was converted from Hermann Oberth's launch tower for the *Frau im Mond* movie demonstration rocket of 1929.

44. Another American, then Midshipman Robert C. Truax (center, uniform), shows his regeneratively-cooled 55-kg (25 lb) thrust motor to British Interplanetary Society, July 1938, at home of Ralph Smith, Chingford, London. Left to right: H. E. Ross, J. H. Edwards, H. E. Turner, Truax, R. A. Smith, M. K. Hanson, and Arthur C. Clarke. Eric Burgess took the photo.

45. Static runs were more typical of VfR and ARS activities. Footage of German film shows successful test of "Aepyornis egg" motor, 286 kgs. (130 lbs) thrust; it resembled giant prehistoric Aepyornis bird egg of Madagascar.

46. Later American static test, Midvale, New Jersey, 1941. This stand, now in National Air and Space Museum, provided invaluable data on Wyld regeneratively-cooled and other motors. Compressed nitrogen bottle for forcing in fuel in motor, in foreground.

47. A positive spin-off from the ARS' static tests, the Wyld regeneratively-cooled rocket motor held here by its inventor James H. Wyld, during 1941 Midvale tests. The motor is also on exhibit in the National Air and Space Museum.

48. Founder of the *Gesellschaft für Weltraumforschung* (Society for Space Research) and later German groups, Hans K. Kaiser, afterwards worked at the Peenemünde rocket center.

49.   Ernst Loebell, Cleveland Rocket Society founder, shows a rocket motor, ca. **1934**

50. Other nations also engaged in astronautics. Alexandre Ananoff, left, succeeded in bringing a French group together in 1938, but it soon disbanded. Ananoff resurrected the movement in France after the war and in 1950 single-handedly created the International Astronautical Federation. He is shown at the 1954 IAF Congress, Innsbruck, Austria, with American delegate and former Assistant Director, Astronautics, National Air and Space Museum, Frederick C. Durant, III.

JOURNAL

OF THE
British Interplanetary Society

JANUARY 1939

6d. to non-members.

Design for a Lunar Space-ship.   *See article page 4.*

51. It was fitting that one decade after the start of the societies, the BIS drew up elaborate plans for a manned lunar mission—a mission realized within some of their life-times.

52. American Rocket Society Proving Stand No. 2 today on display in the National Air and Space Museum.

Illustration Credits

1   Savinien Cryano de Bergerac *Les Oeuvres de Monsieur de Cyrano de Bergerac.* (Amsterdam: Chez J. Desbordes, 1790), frontispiece. (Smithsonian Institution photo A39059).

2   Jules Verne *From the Earth to the Moon . . . and a Trip Around it.* (New York: Scribner, Armstrong & Co., 1874), 296, (SI photo A40863E).

3   Camille Flammarion *La Planète Mars.* (Paris: Gauthier-Villars, 1892), 1, opposite 296. (SI photo 76-14469).

4   H. G. Wells *War of the Worlds.* New York: Harper & Bros., 1898), opposite 78. (SI photo 76-7419).

5   *Science Wonder Quarterly 1,* (Fall 1929): front cover. (SI photo 77-3931).

6   Willy Ley Collection, National Air and Space Museum. (SI photo A3542).

7   *American Weekly,* 13 January 1929. (SI photo 76-17238).

8   National Air and Space Museum. (SI photo 76-7770).

9   National Air and Space Museum. (SI photo 76-13636).

10 National Air and Space Museum. (SI photo A979-B).

11 Werner Brügel *Männer der Rakete.* (Leipzig: Hachmeister & Thall, 1933):33. (SI photo 76-16743).

12 National Air and Space Museum. (SI photo A979-A).

13 National Air and Space Museum. (SI photo A2450).

14 *Tekhnika i Zhin 12* (1924):84. (SI photo 73-1984).

15 Willy Ley Collection, National Air and Space Museum. (SI photo 79-6248).

16 National Air and Space Museum. (SI photo 74-11500).

17 David Lasser, Rancho Bernardo, San Diego, California. (SI photo 84-3408).

18 Herbert Schaefer Collection, National Air and Space Museum. (SI photo 77-4214).

19 Reprinted from *Popular Mechanics, 57* (March 1932):463. ©The Hearst Corporation. All rights reserved. (SI photo 80-10308).

20 Herbert Schaefer Collection, National Air and Space Museum. (SI photo 77-6008).

21 G. Edward Pendray Papers, Seeley Mudd Manuscript Library, Princeton University, Princeton, New Jersey. (SI photo 77-14788).

22 P. E. Cleator, Merseyside, England. (SI photo 78-3943).

23 *Die Rakete 1* (January-June 1927): front page. (SI photo A4315H).

24 *Astronautics 37* (July 1937): front page. (SI photo 77-14800).

25 *Journal of the British Interplanetary Society 1* (January 1934): front page. (SI photo 77-14787).

26 Ernst Loebell, Chagrin Falls, Ohio. (SI photo 79-1746).

27 Herbert Schaefer Collection, National Air and Space Museum. (SI photo 78-17142).

28 Heinz Gartman *The Men Behind the Space Rockets.* (New York: David McKay, 1956): opposite 96. (SI photo 76-7559).

29 Rolf Engel Collection, National Air and Space Museum. (SI photo A3900).

30 G. Edward Pendray Collection, National Air and Space Museum. (SI photo A4555B).

31 G. Edward Pendray Collection, National Air and Space Museum. (SI photo A4555).

32 G. Edward Pendray Collection, National Air and Space Museum. (SI photo A4555E).

33 Rolf Engel Collection, National Air and Space Museum. (SI photo A3913).

34 Pendray Papers. (SI photo 80-1265).

35 G. Edward Pendray Collection, National Air and Space Museum. (SI photo A4556D).

36 Rolf Engel Collection, National Air and Space Museum. (SI photo A3919).

37 K. E. Tsiolkovsky State Museum for Cosmonautics, Kaluga, USSR. (SI photo 83-7149).

38 I. I. Kulagin "Developments in Rocket Engineering Achieved by the Gas Dynamics Laboratory in Leningrad." In Frederick C. Durant, III and George S. James, editors, "First Steps Toward Space," Smithsonian Annals of Flight, 10:92 (Washington, D.C.: Smithsonian Institution Press, 1974). (SI photo 78-3955).

39 K. E. Tsiolkovsky State Museum for Cosmonautics, Kaluga, USSR. (SI photo 73-7133).

40 National Air and Space Museum. (SI photo 8014959).

41 National Air and Space Museum. (SI photo 80-14958).

42 G. Edward Pendray Papers, Seeley Mudd Library, Princeton University, Princeton, New Jersey. (SI photo 78-3949).

43 Reprinted from *Popular Mechanics. 57 (March 1932); 460-61.* ©The Hearst Corporation. All rights reserved. (SI photo 80-14960).

44 National Air and Space Museum. (SI photo 75-11854).

45 Herbert Schaefer Collection, National Air and Space Museum. (SI photo 77-4800).

46 Thiokol Chemical Corporation, Reaction Motors Division, Photo #34-10. (SI photo 75-11488).

47 G. Edward Pendray Collection, National Air and Space Museum. (SI photo A4561).

48 Andrew G. Haley Collection, National Air and Space Museum. (SI photo 80-1270).

49   Ernst Loebell, Chagrin Falls, Ohio. (SI photo 76-1621).

50   Frederick C. Durant, III Collection, National Air and Space Museum. (SI photo 74-154).

51   *Journal of the British Interplanetary Society 5* (January 1939): front page. (SI photo 77-14786).

52   National Air and Space Museum. (SI photo 82-39).

Index